Inferential Statistics

A Contemporary Approach

Richard P. Runyon
C. W. Post College of
Long Island University

Inferential Statistics

A Contemporary Approach

Addison-Wesley Publishing Company

Reading, Massachusetts
Menlo Park, California
London • Amsterdam
Don Mills, Ontario • Sydney

**This book is in the
Addison-Wesley Series in Statistics**

Copyright © 1977 by Addison-Wesley Publishing Company, Inc. Philippines
copyright 1977 by Addison-Wesley Publishing Company, Inc.

ISBN 0-201-06653-X
ABCDEFGHIJ-AL-7987

This book is the second in a series designed to provide a contemporary approach to instruction in statistics. The first book (*Descriptive Statistics: A Contemporary Approach*) covered the essentials of descriptive statistics. The second book advances into the realm of inferential statistics.

The book bears the words "A Contemporary Approach" in its title to emphasize two important themes that are illustrated in the book: (1) the relevance of statistics to a broad spectrum of contemporary disciplines and (2) the possibility of achieving conceptual and computational mastery of the subject matter without a deadening array of complicated formulas and esoteric mathematical proofs.

Inferential Statistics represents many years of cumulative experience in both classroom instruction and textbook construction in the field of statistics. It represents an effort to apply my comprehension of the learning process to a field which many students approach with both awe and apprehension. Several noteworthy features in this respect are:

1. The format permits students to follow the step-by-step solution to a given statistical problem and then apply the knowledge immediately to a series of exercises that appear in the margins of the text. We know that the immediacy of feedback enhances the learning and retention processes.

2. They are then able to check the accuracy of their answers by reference to the worked solutions that appear at the end of the book. In fact, where called for, the solutions are detailed so that the students will be able to locate the source of error whenever their answers disagree with the ones appearing in the text. There are several hundred of these exercises and answers.

3. Review appears repeatedly throughout the book. For example, the concept of a critical value for rejecting the null hypothesis is introduced in Chapter 2. Critical values are subsequently reviewed throughout the book whenever a new sampling distribution is introduced.

4. Because review is built in, individual chapters can stand largely by themselves. This feature makes the book valuable outside the classroom as a reference on computational methods. For example, if a researcher wants to have an assistant run a planned comparison test on a set of means, he or she may direct the assistant's attention to Chapter 9, where the Tukey HSD test is presented.

5. In my experience, most students will encounter no difficulty following the procedures necessary to calculate a test statistic. However, some will stub their toes when they must subsequently refer to a statistical table. For this reason, I have taken care to see that the student is instructed in the use of the various tables. In fact, the use of a given table is frequently introduced *prior* to the discussion of the test statistic to which it applies.

There are a variety of ways in which the book can be used. It can serve by itself as the text for a second course in statistics or, in conjunction with *Descriptive Statistics: A Contemporary Approach*, it can serve as the text for a full semester course in statistics.

Preface

Because of the "building-block" nature of the book's development, it is ideally suited for self-paced courses that are appearing with increasing frequency on the academic scene. I am referring to courses variously referred to as Personalized Instruction, PSI, and the like. The test booklet will be particularly useful in such courses. Each chapter of the booklet includes two detachable alternate forms of test instruments.

Inferential Statistics : A Contemporary Approach may also be used as a supplementary text in courses for which a regular text has been assigned. Altogether too often, in my experience, students falter because their inability to calculate a given statistic produces negative side effects that interfere with conceptual processes. When this book is used as a supplementary text, the instructor may breathe easy in this respect. He or she may then feel free to devote class time to the important cognitive aspects of statistical analysis. This text integrates well with most books on the market since it covers most topics with a minimum of computational formulas and esoteric symbols.

I wish to thank my favorite co-author, Audrey Haber, for permission to reproduce some materials appearing in our book *Fundamentals of Behavioral Statistics*, 3rd Edition, and *General Statistics*, 3rd Edition. I would also like to express my deepest appreciation to Mrs. Robin Schafer for her good cheer while performing a task that must have been painful at times.

R. P. R.

Tucson, Arizona
January 1977

Contents

1 Introduction to Inferential Statistics

Probability is expressed as a proportion   2
The standard normal curve and probability   3
The concept of a sampling distribution   5

2 Testing Statistical Hypotheses

Testing statistical hypotheses: level of significance   13
Testing statistical hypotheses: null hypothesis and alternative
hypothesis   14

3 Tests of Significance, One-Sample Case

Testing hypotheses about the mean when parameters are known   22
Testing hypotheses about the mean when parameters are
unknown   24
Test of significance for Pearson r, one-sample case   30

**4 Significance of Difference Between Means,
Independent Samples**

Test of significance when parameters are known   36
Test of significance when parameters are unknown   39
The test for homogeneity of variance   45
Estimating degree of association between the experimental
and dependent variables   46

**5 Significance of Difference Between Pearson r's,
Independent Samples**

Transforming r to z_r 52
Calculating the standard error of the difference between
z's (s_{D_z}) 52
Applying z as the test statistic   53

**6 Significance of Difference Between Means,
Correlated Samples**

The main classes of correlated samples design   58
The Student t-ratio, correlated samples   58
The Sandler A-statistic as an alternative to Student's t-ratio   63

7 Power of a Statistical Test

The two types of errors   68
Calculating the probability of a type II or type β error   69
The concept of power   72
Improving the power of a test   72

8   One-Way Analysis of Variance, Independent Samples

Between-group and within-group variance estimates   78
Partitioning the sum of squares   79
Obtaining variance estimates   82
Raw score formulas for total, between-group, and within-group
sum of squares   83

9   Multicomparison Tests

Planned (*a priori*) vs. unplanned (*a posteriori*) comparisons   90
The Tukey HSD (honestly significant difference) test   90

10   One-Way Analysis of Variance, Correlated Samples Without Repeated Measures (Randomized Block Design)

The main classes of correlated samples design   96
Partitioning the sum of squares   97

11   Two-Way Analysis of Variance, Factorial Design Employing Independent Samples

The concept of a treatment combination   106
Partitioning the sum of squares   107
Applying the Tukey HSD test to differences among B-variable
means   116

Appendix

Table A   Percent of area under the standard normal curve   123
Table B   Critical values of t 137
Table C   Critical values of F 138
Table C_1 Values of F exceeded by 0.025 of the values in the
           sampling distribution   141
Table D   Critical values of A 142
Table E   Percentage points of the Studentized range   143
Table F   Transformation of r to z_r 144
Table G   Functions of r 145

Answers   148

Index   173

Introduction to Inferential Statistics

In descriptive statistics you learned how to make sense out of a mass of raw data. You learned how to calculate and interpret descriptive statistics for describing both the central tendency of a distribution of scores or quantities (mean, median, and mode) and the dispersion of scores about central tendency (range, average deviation, standard deviation, and variance). In brief, your focus in descriptive statistics was to describe, with both accuracy and an economy of statement, aspects of samples that were selected from certain populations.

In inferential statistics, your focus shifts from near vision to far vision. You shift from describing samples to making inferences about populations. However, descriptive statistics remains important since it provides the factual base for "taking the inductive leap" from samples to populations.

Probability theory provides both the techniques and the rationale for engaging in the inferential function. In this chapter, we shall present a sufficient introduction to probability theory to permit you to master the concepts of inferential statistics presented in the book. The treatment of probability theory is selective rather than comprehensive.

PROBABILITY IS EXPRESSED AS A PROPORTION

Most commonly, probability is expressed as a proportion. The proportion may vary from 0.00 to 1.00. When $p = 0.00$, the probability is zero that the event in question will occur; on the other hand, when $p = 1.00$, the event is certain to occur.

Examples

If $p = 0.50$, the chances are 50 in 100 (or 1 in 2) that the event will occur.

If $p = 0.95$, the chances are 95 in 100 that the event will occur.

If $p = 0.25$, the likelihood is 25 in 100 (or 1 in 4) that the event will occur.

Do Exercises 1 through 4 at the right.

Probability may be expressed as a proportion representing the number of events favoring a given outcome relative to the total number of events possible:

$$p_A = \frac{\text{number of events favoring } A}{\text{number of events favoring } A + \text{number of events not favoring } A}$$

OBJECTIVES

1. Know the way in which probability is expressed.

2. Know how to use the standard normal curve to calculate the probabilities of various events.

3. Know the distinction between one- and two-tailed probability values.

4. Understand the concept of a sampling distribution.

5. Know how to construct a sampling distribution of means for $n = 2$ when the population is an extremely limited number of scores.

6. Know how to use the sampling distribution to calculate the probability values of various events.

EXERCISES

1. Express verbally the meaning of $p = 0.05$.

2. Express verbally the meaning of $p = 0.01$.

3. If the chances are 40 in 100 that the event will occur, $p = $ _____ .

4. If the chances are 75 in 100 that the event will occur, $p = $ _____ .

Examples

If we define a head as event A and a tail as event non-A, the probability of a head on a single toss of a coin is

$$p_A = \frac{1}{1 + 1} = \frac{1}{2}$$

If we define a two as event A, and a one, three, four, five, and six each as event non-A, the probability of a two on a single toss of a die is

$$p_A = \frac{1}{1 + 5} = \frac{1}{6}.$$

Do Exercises 5 through 10 at the right.

Probability may also be expressed as the proportion of one area under a curve relative to the total area under the curve.

In descriptive statistics, we discussed z-scores and areas under the standard normal curve. We saw that for a given z there is a corresponding percentage of area both above and below that z. Areas under the standard normal curve are shown in Table A. If we move the decimal two places to the left in columns A, B, and C of Table A, the cell entries represent proportions of total area under the curve.

Examples

What is the proportion of area beyond $z = -1.07$ (i.e., $z \leq -1.07$)? In column C we find 14.23. Moving the decimal two places to the left, we see that the proportion of area beyond $z = -1.07$ is 0.1423.

What is the proportion of area above $z = -1.07$? Since the total area under the normal curve is 1.00, we may obtain the answer by subtraction: $1.0000 - 0.1423 = 0.8577$.

Do Exercises 11 through 24 at the right.

THE STANDARD NORMAL CURVE AND PROBABILITY

Recall that the standard normal curve has a mean equal to zero, a standard deviation equal to 1.00, and a total area under the curve equal to 1.00. By transforming scores of normally distributed variables into z-scores, we may determine the probability of various events.

In a single draw from a 52-card deck of playing cards, what is the probability of selecting:

5. The ace of spades?

6. Any ace?

7. A club?

8. A black card?

9. A red nine?

10. A card higher than a ten (ace is considered a high card)?

Find the proportion of the area under the standard normal curve of the following z's.

11. Area above $z = 1.53$.

12. Area below $z = 1.53$.

13. Area above $z = -1.74$.

14. Area below $z = -1.74$.

15. Area above $z = 0.00$.

16. Area below $z = 0.00$.

17. Area above $z = 2.63$.

18. Area below $z = -2.05$.

19. Area above $z = 1.96$.

20. Area below $z = -1.96$.

21. Area above $z = 2.33$.

22. Area below $z = -2.33$.

23. Area above $z = -2.33$.

24. Area below $z = 2.33$.

Examples

a) Given a normally distributed variable with a mean equal to 500 and a standard deviation equal to 100, what is the probability that, drawing a score at random, we will select a score equal to 600 or more? We first calculate z:

$$z = \frac{600 - 500}{100} = 1.00.$$

From Table A we see that the proportion of area equal to and beyond $z = 1.00$ is 0.1587. Thus the probability of selecting a score equal to or greater than 600 is 0.1587.

b) Given the above, what is the probability of selecting a score equal to or less than 600?

$$P_{z \leq 1.00} = 1.0000 - 0.1587$$
$$= 0.8413$$

c) Given that the mean (μ) of a normally distributed variable is 50 and the standard deviation (σ) is 3. What is the probability of selecting a score equal to or less than 45?

$$z = \frac{45 - 50}{3} = -\frac{5}{3} = -1.67$$

$$P_{z \leq -1.67} = 0.0475$$

Do Exercises 25 through 29 at the right.

One- and Two-Tailed Probability Values

In all of the above examples, the probability was obtained by looking at only one end of the distribution. We asked such questions as: What is the probability of obtaining a score equal to or greater than or equal to or less than some specific value? Very often in inferential statistics, we want to express the probabilities in terms of both extremes of a distribution: What is the probability of obtaining a score as rare as or as deviant from the mean as some specified value?

Examples

a) In a normal distribution with a mean equal to 100 and a standard deviation equal to 10, what is the probability of obtaining a score as unusual as 115?

Since a normal distribution is symmetrical about the mean, a score of 85 is as deviant as a score of 115. All we need do is find

Given $\mu = 100$, $\sigma = 15$, find the following probability values.

25. The probability of selecting a score equal to or greater than 122.

26. The probability of selecting a score equal to or greater than 89.

27. The probability of selecting a score equal to or less than 89.

28. The probability of selecting a score equal to or greater than 103.

29. The probability of selecting a score equal to or less than 103.

the probability of obtaining a score equal to or greater than 115 and then double the *p*-value in order to take into account both tails of the distribution.

$$z_{x=115} = \frac{115 - 100}{10}$$

$$= 1.50$$

$$P_{x \geq 115} = 0.0668$$

The two-tailed probability value is $2 \times 0.0668 = 0.1336$.

b)   In a normal distribution with $\mu = 500$ and $\sigma = 100$, find the probability of selecting, by chance, a score as deviant as 280.

$$z_{x=280} = \frac{280 - 500}{100}$$

$$= -2.20$$

$$P_{z \leq -2.20} = 0.0139$$

The two-tailed probability value is $2 \times 0.0139 = 0.0278$.

Do Exercises 30 through 35 at the right.

THE CONCEPT OF A SAMPLING DISTRIBUTION

One of the key concepts in inferential statistics is that of a sampling distribution. *A sampling distribution is a theoretical probability distribution of the possible values of some sample statistic that would occur if we were to draw all possible samples of a fixed size from a given population.*

Sampling distributions provide the theoretical probability values with which we can evaluate statements about various sample statistics. For example, if we know the precise form that a distribution of samples means, $n = 4$, will take when drawn from a given population, we are in a position to assess the probabilities that a sample mean based on $n = 4$ was drawn from that population.

The procedures are quite analogous to the use of the standard normal curve to make probability statements about scores. The standard normal curve may be regarded as a sampling distribution for evaluating samples of size $n = 1$ (single scores).

Example

Imagine a population consisting of the following five scores: 0, 1, 2, 3, 4. We construct a theoretical distribution of all possible sample

Find the two-tailed probability values of the following scores selected at random from a normally distributed variable.

30. $\mu = 50, \sigma = 5, X = 57$.

31. $\mu = 100, \sigma = 16, X = 82$.

32. $\mu = 500, \sigma = 100, X = 730$.

33. $\mu = 40, \sigma = 10, X = 51$.

34. $\mu = 500, \sigma = 100, X = 270$.

35. $\mu = 100, \sigma = 16, X = 143$.

means of $n = 2$ drawn from this population. We sample *with replacement*—i.e., for each sample of $n = 2$, we select a score, record it, return it, and then select the second score. We find the mean of the pair of scores. The following table shows the results of this sampling experiment. The marginal values show the scores selected on the first and second draw for each sample of $n = 2$. The means are shown in the body of the table.

Second draw	First draw				
	0	**1**	**2**	**3**	**4**
0	0.0	0.5	1.0	1.5	2.0
1	0.5	1.0	1.5	2.0	2.5
2	1.0	1.5	2.0	2.5	3.0
3	1.5	2.0	2.5	3.0	3.5
4	2.0	2.5	3.0	3.5	4.0

Do Exercise 36 at the right.

We can now obtain a frequency distribution of these 25 means.

$\bar{X}$	f
4.0	1
3.5	2
3.0	3
2.5	4
2.0	5
1.5	4
1.0	3
0.5	2
0.0	1
	$N_{\bar{X}} = 25$

In this table, $N_{\bar{X}}$ indicates the number of means in the sampling distribution.

Do Exercise 37 at the right.

By dividing each frequency by the total number of sample means ($N_{\bar{X}} = 25$), we obtain a probability distribution of means.

36. A given population consists of the following seven scores: 1, 2, 3, 4, 5, 6, 7. Draw up a table showing the sample means obtained by selecting, with replacement, all possible samples of $n = 2$.

37. Construct a frequency distribution of the 49 sample means obtained in Exercise 36.

$\bar{X}$	f	$p_{\bar{X}}$	$\bar{X}$	f	$p_{\bar{X}}$
4.0	1	0.04	1.5	4	0.16
3.5	2	0.08	1.0	3	0.12
3.0	3	0.12	0.5	2	0.08
2.5	4	0.16	0.0	1	0.04
2.0	5	0.20			
					$\sum p_{\bar{X}} = 1.00$

The resulting probability distribution is the sampling distribution of the mean for samples of $n = 2$ drawn from the indicated population. Note that $\sum p_X = 1.00$.

Do Exercise 38 at the right.

We are now in a position to raise questions concerning sample means of $n = 2$ drawn from the indicated population.

Examples

a) What is the probability of drawing a mean equal to 4.0?

$p_{\bar{X}=4.0} = 0.04$

b) What is the probability of drawing a mean as deviant or as unusual as 4.0? Since a sample mean of 0.0 is equally deviant as a mean of 4.0, we merely double the above p-value to obtain

$p_{\bar{X}=4.0 \text{ or } 0.0} = 0.08.$

Alternatively, we add together the two answers:

$p_{\bar{X}=4.0 \text{ or } 0.0} = p_{\bar{X}=4.0} + p_{\bar{X}=0.0}$

$\qquad = 0.04 + 0.04$

$\qquad = 0.08.$

c) What is the probability of drawing a mean equal to 3.0 or larger?

$p_{\bar{X} \geq 3.0} = p_{\bar{X}=3.0} + p_{\bar{X}=3.5} + p_{\bar{X}=4.0}$

$\qquad = 0.12 + 0.08 + 0.04$

$\qquad = 0.24$

Do Exercises 39 through 43 at the right.

In Chapter 2, we shall show how a sampling distribution may be used to explore or test statistical hypotheses.

38. Construct a probability distribution of the means from the frequency distribution in Exercise 37. Verify that $\sum p_{\bar{X}} = 1.00$.

Answer the following questions by using the sampling distribution of the mean, $n = 2$, constructed in Exercise 38.

39. What is the probability of selecting a mean equal to or greater than 7.0?

40. What is the probability of selecting a mean equal to or greater than 6.5?

41. What is the probability of selecting a mean as deviant as 6.5?

42. What is the probability of selecting a mean as deviant as 2.0?

43. What is the probability of selecting a mean equal to or less than 3.0?

CHAPTER 1 TEST

To complete this test, students will require access to Table A (Percentage of Areas under the Standard Normal Curve).

1. If a person calculated a p-value equal to 1.56, you would know:

    a)   the probability of the event is very high.

    b)   an error has been made.

    c)   the probability of the event is very low.

    d)   the probability of the event is about 1.5 in 2.

2. A p-value equal to 0.33 means that the event:

    a)   has about 33 chances in 50 of occurring.

    b)   is very unlikely to occur.

    c)   has about 1 chance in 3 of occurring.

    d)   has about 3 chances in 1000 of occurring.

3. If an event has 95 chances out of 100 of occurring, the associated probability is:

    a)   95.00   b)   0.0095   c)   0.0950   d)   0.9500

4. If $p = 0.63$, the chances that the event will occur are:

    a)   63 in 100

    b)   63 in 1000

    c)   almost 1 in 100

    d)   almost certain

5. If the number of events favoring A is 40 and the number favoring non-A is 80, p_A equals:

    a)   0.50   b)   0.67   c)   0.33   d)   0.05

6. If the number of events favoring A is 60 and the number favoring non-A is 40, p_A equals:

    a)   0.60   b)   0.40   c)   1.50   d)   0.67

7. If the number of events favoring A is 65 and the number favoring non-A is 55, $p_{\text{non-}A}$ equals:

    a)   1.18   b)   0.85   c)   0.55   d)   0.46

8. If $z = -2.17$, the proportion of area beyond z is:

    a)   0.0150   b)   0.9850   c)   0.0300   d)   01.50

9. Given a normally distributed variable with a mean equal to 50 and a standard deviation equal to 3, what is the probability of drawing a score at random equal to or less than 43?

    a) 00.99   b) 0.0264   c) 0.0099   d) 98.98

10. Given a normally distributed variable with a mean equal to 500 and a standard deviation equal to 100, what is the probability of drawing a score at random equal to or less than 250?

    a) 0.9938   b) 0.0062   c) 0.6200   d) 0.4938

11. Given a normally distributed variable with a mean equal to 100 and a standard deviation equal to 15, what is the probability of drawing a score as rare as 126?

    a) 0.0418   b) 0.0836   c) 0.4882   d) 0.9154

Questions 12–15 are based on the following sampling distribution of the mean.

$\bar{X}$	$p_{\bar{X}}$
10.00	0.0039
9.25	0.0156
8.50	0.0391
7.75	0.0781
7.00	0.1211
6.25	0.1562
5.50	0.1719
4.75	0.1562
4.00	0.1211
3.25	0.0781
2.50	0.0391
1.75	0.0156
1.00	0.0039

12. What is the probability of selecting a mean equal to or less than 4.00?

    a) 0.1211   b) 0.2422   c) 0.2578   d) 0.5156

13. What is the probability of selecting a mean equal to or greater than 8.50?

    a) 0.0586   b) 0.0391   c) 0.1172   d) 0.0782

14. What is the probability of selecting a mean as unusual as 9.25?

    a) 0.0196   b) 0.0312   c) 0.0390   d) 0.0195

15. What is the probability of selecting a mean as unusual as 3.25?

    a) 0.1367   b) 0.0781   c) 0.1562   d) 0.2734

Practical

16. Given a normally distributed variable with $\mu = 50$ and $\sigma = 3$, find the probability that:

    a)   a score selected at random will be equal to or greater than 55

    b)   a score selected at random will be equal to or less than 46

    c)   a score selected at random will be equal to or greater than 57

    d)   a score selected at random will be equal to or less than 41

    e)   a score selected at random will be equal to or less than 42

    f)   a score selected at random will be equal to or greater than 51

    g)   a score selected at random will be equal to or greater than 59

    h)   a score selected at random will be equal to or less than 48.

    i)   a score selected at random will be as unusual as 55

    j)   a score selected at random will be as unusual as 44

    k)   a score selected at random will be as unusual as 41

    l)   a score selected at random will be as unusual as 59

    m)   a score selected at random will be as unusual as 52

17. Given the following nine scores: 0, 1, 2, 3, 4, 5, 6, 7, 8. Construct a sampling distribution using a sample size of $n = 2$, sampling with replacement. Answer the following questions.

    a)   What is the probability of selecting at random a mean equal to or greater than 7.5?

    b)   What is the probability of selecting at random a mean as rare as 0?

    c)   What is the probability of selecting at random a mean as high as 6.5?

    d)   What is the probability of selecting at random a mean as unusual as 0.5?

    e)   What is the probability of selecting at random a mean as low as 2.0?

Testing Statistical Hypotheses

In Chapter 1 you learned a few basic concepts of probability theory and the importance of sampling distributions in inferential statistics. In this chapter we shall explore the use of sampling distributions in testing statistical hypotheses.

In Chapter 1, you drew all possible samples of $n = 2$ from a population of seven scores. You subsequently constructed a sampling distribution of means for $n = 2$ and you answered a number of questions regarding the probability of obtaining certain events (means) when drawing samples of $n = 2$ with replacement from that population.

The following sampling distribution of means is based upon drawing all possible samples of $n = 2$ with replacement from a population of seven scores.

$\bar{X}$	f	$p_{\bar{X}}$
7.0	1	0.0204
6.5	2	0.0408
6.0	3	0.0612
5.5	4	0.0816
5.0	5	0.1020
4.5	6	0.1224
4.0	7	0.1429
3.5	6	0.1224
3.0	5	0.1020
2.5	4	0.0816
2.0	3	0.0612
1.5	2	0.0408
1.0	1	0.0204
$N_{\bar{X}} = 49$		$\sum p_{\bar{X}} = 0.9997*$

* The disparity of 0.0003 represents rounding error.

Let us now change the problem somewhat. Imagine that you draw samples of $n = 2$ from some *unknown* population. Your task is to decide whether or not it is *likely* that the mean was drawn from the population from which the above sampling distribution was constructed.

To aid in the decision making process, we shall make an arbitrary rule:

a) If the event in question or an event more rare would have occurred 5% of the time or less, we shall decide that the mean was *not* drawn from the population of seven scores; we shall call this the *reject* decision.

b) If the event in question or events more rare would have occurred more than 5% of the time, we shall entertain as possible the hypothesis that the mean was drawn from the indicated population; we shall call this the *fail to reject* decision.

Note that the *fail to reject* decision does not lead to the positive assertion that the event was drawn from the known population. It merely acknowledges the reasonable possibility that it was.

Examples

a) Is it likely that a mean equal to or less than 2.0 was drawn from the population of seven scores?

$$p_{\bar{X} \leq 2.0} = p_{\bar{X}=2} + p_{\bar{X}=1.5} + p_{\bar{X}=1.0}$$
$$= 0.0612 + 0.0408 + 0.0204$$
$$= 0.1224$$

Since $p_{\bar{X} \leq 2.0}$ is greater than 0.05, we make the *fail to reject* decision.

b) Is it likely that a mean equal to or less than 1.0 was drawn from the population of seven scores?

$$p_{\bar{X} \leq 1.0} = p_{\bar{X}=1.0}$$
$$= 0.0204$$

Since $p_{\bar{X} \leq 1.0}$ is less than 0.05, we make the *reject* decision. In accordance with our rule, we conclude it was unlikely that the mean of 1.0 was drawn from a population of seven scores in which $\mu = 4.0$.

Do Exercises 1 through 6 at the right.

TESTING STATISTICAL HYPOTHESES: LEVEL OF SIGNIFICANCE

Knowing the sampling distribution of a statistic permits us to label certain events as common, others as somewhat unusual, and still others as rare. Probability theory provides the basis for these judgments. Moreover, if we conclude that the occurrence of an event (for example, the outcome of an experiment) is rare, we conclude that nonchance factors (e.g., the effect of an experimental variable) are responsible for or have caused this unusual or rare event. Again, probabilities provide the basis for this inference.

In the preceding section, we were actually employing one of the conventional cutoff points for inferring the operation of nonchance factors (that is, making the reject decision). It is referred to as the *0.05 significance level*. When the reject decision has been made at the 0.05 level, it is conventional to refer to the outcome of the experiment as statistically significant at the 0.05 level.

EXERCISES

Use the sampling distribution of the mean in the text at the left to make the *reject* or *fail to reject* decision with regard to the following questions.

1. Is it likely that a mean as rare as 1.0 was drawn from the population of seven scores?

2. Is it likely that a mean equal to or greater than 7.0 was drawn from the indicated population?

3. Is it likely that a mean as rare as 6.5 was drawn from the indicated population?

4. Is it likely that a mean as rare as 7.0 was drawn from the indicated population?

5. Is it likely that a mean of 1.5 or less was drawn from the indicated population?

6. Is it likely that a mean of 5.5 or greater was drawn from the indicated population?

When the event or one that is more deviant would occur 5% of the time or less, by chance, we are willing to take the inductive leap and assert that the results are due to nonchance factors.

Other scientists prefer to use the *0.01 significance level* or the *1% significance level*: When the event or one that is more deviant would occur 1% of the time or less, by chance, we assert that the results are due to nonchance factors. It is then conventional to refer to the results of the experiment as statistically significant at the 0.01 level.

Examples

a) If a researcher is using the 0.05 significance level and the probability associated with the outcome of an experiment is 0.03, he asserts the operation of nonchance factors.

b) For the same outcome, if the researcher were using the 0.01 significance level, he would not assert the operation of nonchance factors. He would make the *fail to reject* decision.

Do Exercises 7 through 14 at the right.

The level of significance set by the researcher for making the *reject decision* is known as the *alpha* (α) *level*. When we employ the 0.01 significance level, $\alpha = 0.01$. When we employ the 0.05 significance level, $\alpha = 0.05$.

Do Exercises 15 and 16 at the right.

TESTING STATISTICAL HYPOTHESES: NULL HYPOTHESIS AND ALTERNATIVE HYPOTHESIS

Prior to the beginning of an experiment, the researcher sets up two mutually exclusive hypotheses. One is a statistical hypothesis which the experimenter generally expects to disprove. It is referred to as the *null hypothesis* (H_0). The null hypothesis states some expectation concerning the value of one or more population parameters—e.g., the mean of the population is some specific value, such as 100; or the samples were drawn from the same population of means so that $\mu_1 - \mu_2 = 0$.

Examples

a) If we were testing the honesty of a coin, the null hypothesis would read

H_0: the coin is unbiased, that is, $P = Q = \frac{1}{2}$,

In the following exercises, we give the level of significance used by the researcher and the *p*-value associated with the outcome of an experiment. State whether you would make the *reject* or *fail to reject* decision.

7. Level of significance, 0.05; $p = 0.65$.

8. Level of significance, 0.01; $p = 0.009$.

9. Level of significance, 0.01; $p = 0.025$.

10. Level of significance, 0.05; $p = 0.025$.

11. Level of significance, 0.01; $p = 0.10$.

12. Level of significance, 0.05; $p = 0.49$.

13. Level of significance, 0.05; $p = 0.049$.

14. Level of significance, 0.01; $p = 0.012$.

15. When a researcher uses the 0.01 significance level, _____ = _____ .

16. When a researcher uses the 0.05 significance level, _____ = _____ .

where

P = the probability of a head and Q = the probability of a tail.

b) A drug is administered to one group (X_1) and a placebo to a second (X_2). The null hypothesis is

H_0: the drug does not effect the dependent variable, that is, $\mu_1 = \mu_2$ or $\mu_1 - \mu_2 = 0$,

where μ_1 is the population represented by the X_1 sample and μ_2 is the population represented by the X_2 sample.

Do Exercises 17 and 18 at the right.

The *alternative hypothesis* (H_1) in effect denies the null hypothesis. If the null hypothesis states that there is no difference in the population means from which the two samples were drawn, the alternative hypothesis asserts that there is a difference. The alternative hypothesis usually states the researcher's expectations.

Examples

In (a), above:

H_1: the coin is biased, that is, $P \neq Q \neq \frac{1}{2}$.

In (b), above:

H_1: the drug affects the dependent variable, that is, $\mu_1 \neq \mu_2$ or $\mu_1 - \mu_2 \neq 0$.

Do Exercises 19 and 20 at the right.

Directional and Nondirectional Hypotheses

Occasionally, a researcher will have developed a sufficiently detailed theory or amassed enough evidence to permit him or her to predict, in the alternative hypothesis, the direction of the difference. The alternative hypothesis then states the expected direction of the difference.

Example

A researcher hypothesizes that a specific drug leads to slower reaction times on a psychomotor task. Group X_1 receives the drug and X_2 the placebo. The alternative hypothesis would be

H_1: $\mu_1 < \mu_2$, that is, the population mean of the drug group is less than the population mean of the placebo group.

17. A researcher wishes to determine the effect of the number of hours of study on academic performance. One group of subjects is given 8 hours of study per week and a second group is given only 4 hours a week. Set up the null hypothesis.

18. A new aircraft is being designed in which two types of dials are being considered for use on the display panel. One is circular and the other is rectangular. Two groups of subjects are tested on the speed with which they can read the values on each dial. State the null hypothesis.

19. State H_1 for Exercise 17.

20. State H_1 for Exercise 18.

If the alternative hypothesis is directional, so also is the null hypothesis. In the above example, the null hypothesis would read

$H_0: \mu_1 \geq \mu_2$, that is, the population mean of the drug group is equal to or greater than the population mean of the placebo group.

Do Exercises 21 and 22 at the right.

Directional alternative hypotheses are evaluated by one-tailed tests of significance. Nondirectional alternative hypotheses are evaluated by two-tailed tests of significance.

Examples

a) If $\alpha = 0.05$, two-tailed test and the one-tailed p-value of the event in question is 0.05, we fail to reject H_0 since the two-tailed p-value is 0.10, which is greater than $\alpha = 0.05$.

b) If $\alpha = 0.05$, one-tailed test and the one-tailed p-value of the event in question is 0.05, we reject H_0 since the one-tailed p-value equals α. Since we have rejected H_0, we assert H_1.

c) If $\alpha = 0.01$, two-tailed test and the one-tailed p-value of the event in question is 0.004, we reject H_0 since the two-tailed p-value is 0.008, which is less than α. Since we have rejected H_0, we assert H_1.

Do Exercises 23 through 30 at the right.

The Concept of the Critical Value

Many tables that are used by researchers and statisticians do not provide the probability values for the sampling distributions to which they refer. Rather, they present *critical values* which define the *region of rejection* at various levels of α. The region of rejection is *that portion of area under a curve which includes those values of a statistic which lead to rejection of the null hypothesis.*

To illustrate, the four figures on the opposite page show the regions for rejection of H_0 when the standard normal curve is used as a sampling distribution.

Examples

a) If H_0 is $\mu_1 = \mu_2$, H_1 is $\mu_1 \neq \mu_2$, $\alpha = 0.05$, and $z \geq 1.96$ or ≤ -1.96, reject H_0.

b) If H_0 is $\mu_1 \leq \mu_2$, H_1 is $\mu_1 > \mu_2$, $\alpha = 0.05$, and $z \geq 1.65$, reject H_0.

c) If H_0 is $\mu_1 \geq \mu_2$, H is $\mu_1 < \mu_2$, $\alpha = 0.05$, and $z \leq -1.65$, reject H_0.

Do Exercises 31 through 40 at the right.

21. State H_0 and H_1 for Exercise 17. The experimenter expects the mean of the 8-hour group to be higher.

22. State H_0 and H_1 for Exercise 18. The experimenter expects the rectangular dial to lead to higher speed scores.

Given α and the p-values associated with the outcome of an experiment, make the decision whether or not to reject H_0.

23. $\alpha = 0.01$, two-tailed; $p = 0.007$, one-tailed.

24. $\alpha = 0.01$, one-tailed; $p = 0.009$, one-tailed.

25. $\alpha = 0.05$, two-tailed; $p = 0.03$, one-tailed.

26. $\alpha = 0.05$, two-tailed; $p = 0.04$, two-tailed.

27. $\alpha = 0.01$, one-tailed; $p = 0.01$, one-tailed.

28. $\alpha = 0.01$, two-tailed; $p = 0.02$, one-tailed.

29. $\alpha = 0.05$, one-tailed; $p = 0.06$, one-tailed.

30. $\alpha = 0.01$, two-tailed; $p = 0.008$, two-tailed.

2

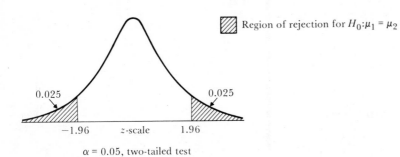

α = 0.05, two-tailed test

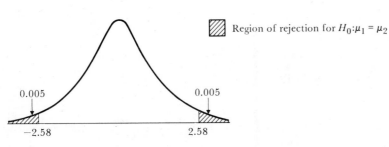

α = 0.01, two-tailed test

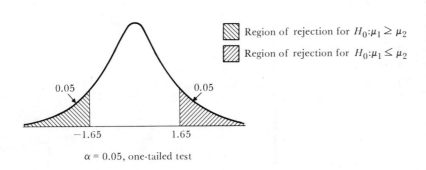

α = 0.05, one-tailed test

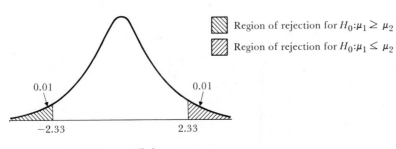

α = 0.01, one-tailed test

In Exercises 31 through 33, H_0, H_1, and α are given. Indicate the value of z defining the region of rejection.

31. $H_0: \mu_1 = \mu_2$, $H_1: \mu_1 \neq \mu_2$, $\alpha = 0.01$.

32. $H_0: \mu_1 \leq \mu_2$, $H_1: \mu_1 > \mu_2$, $\alpha = 0.01$.

33. $H_0: \mu_1 \geq \mu_2$, $H_1: \mu_1 < \mu_2$, $\alpha = 0.05$.

In Exercises 34 through 40, H_0, H_1, α, and z are given. Indicate whether or not the z-statistic lies within the critical region.

34. $H_0: \mu_1 = \mu_2$, $H_1: \mu_1 \neq \mu_2$, $\alpha = 0.05$, $z = -2.02$.

35. $H_0: \mu_1 \geq \mu_2$, $H_1: \mu_1 < \mu_2$, $\alpha = 0.05$, $z = -1.74$.

36. $H_0: \mu_1 = \mu_2$, $H_1: \mu_1 \neq \mu_2$, $\alpha = 0.05$, $z = 2.63$.

37. $H_0: \mu_1 \leq \mu_2$, $H_1: \mu_1 > \mu_2$, $\alpha = 0.01$, $z = 2.21$.

38. $H_0: \mu_1 = \mu_2$, $H_1: \mu_1 \neq \mu_2$, $\alpha = 0.05$, $z = 0.58$.

39. $H_0: \mu_1 \leq \mu_2$, $H_1: \mu_1 > \mu_2$, $\alpha = 0.05$, $z = 1.57$.

40. $H_0: \mu_1 = \mu_2$, $H_1: \mu_1 \neq \mu_2$, $\alpha = 0.01$, $z = 2.34$.

CHAPTER 2 TEST

1. If $\alpha = 0.01$, one-tailed test, which of the following one-tailed p-values would lead to the rejection of H_0?

    a) $p = 0.004$ b) $p = 0.02$ c) $p = 0.05$ d) $p = 0.015$

2. If $\alpha = 0.05$, two-tailed test, which of the following two-tailed p-values would result in the rejection of H_0?

    a) $p = 0.03$ b) $p = 0.058$ c) $p = 0.50$ d) $p = 0.10$

3. If $\alpha = 0.05$, two-tailed test, which of the following one-tailed p-values would lead to the rejection of H_0?

    a) $p = 0.05$ b) $p = 0.02$ c) $p = 0.10$ d) $p = 0.50$

4. If $\alpha = 0.01$, one-tailed test, which of the following two-tailed p-values would lead to the rejection of H_0?

    a) $p = 0.10$ b) $p = 0.20$ c) $p = 0.05$ d) $p = 0.018$

5. If $\alpha = 0.01$, one-tailed test, which of the following two-tailed p-values would lead to the rejection of H_0?

    a) $p = 0.01$ b) $p = 0.02$

    c) $p = 0.026$ d) none of the preceding

6. If $\alpha = 0.05$, one-tailed test, which of the following two-tailed p-values would lead to the rejection of H_0?

    a) $p = 0.20$ b) $p = 0.14$ c) $p = 0.10$ d) $p = 0.50$

Questions 7–10 are based on the following sampling distribution, which was constructed by drawing samples of $n = 4$, with replacement, from the following population of scores: 0, 1, 2, 3; $\mu = 1.50$.

$\bar{X}$	f	p_X
3.00	1	0.0039
2.75	4	0.0156
2.50	10	0.0390
2.25	20	0.0781
2.00	31	0.1211
1.75	40	0.1562
1.50	44	0.1719
1.25	40	0.1562
1.00	31	0.1211
0.75	20	0.0781
0.50	10	0.0390
0.25	4	0.0156
0.00	1	0.0039

7. The critical region for testing $H_0: \mu_1 = 1.50$ at $\alpha = 0.01$ is:

   a) $\bar{X} \leq 0.00$ or $\bar{X} \geq 3.00$

   b) $\bar{X} \geq 0.00$

   c) $\bar{X} \leq 0.25$ or $\bar{X} \geq 2.75$

   d) There is no sample statistic in the sampling distribution sufficiently extreme to test H_0.

8. The critical region for testing $H_0: \mu_1 \leq 1.50$ at $\alpha = 0.05$ is:

   a) $\bar{X} \geq 2.50$ b) $\bar{X} \leq 0.50$ c) $\bar{X} \geq 2.75$ d) $\bar{X} \leq 0.25$

9. The critical region for testing $H_0: \mu_1 \geq 1.50$ at $\alpha = 0.01$ is:

   a) $\bar{X} \geq 0.25$ b) $\bar{X} \leq 0.00$ c) $\bar{X} \geq 1.00$ d) $\bar{X} \leq 1.33$

10. The critical region for testing $H_0: \mu_1 = 1.50$ at $\alpha = 0.05$ is:

    a) $\bar{X} \geq 2.50$ or $\bar{X} \leq 0.50$ b) $\bar{X} \leq 2.50$

    c) $\bar{X} \leq 2.75$ d) $\bar{X} \geq 2.75$ or $\bar{X} \leq 0.25$

Practical

11. Construct a sampling distribution for sample size $n = 2$ from the following population of 13 scores: 0, 1, 2, 3, 4, 5, 6, 7, 8, 9, 10, 11, 12.

12. Imagine that you draw samples of $n = 2$ from an unknown population. Using the sampling distribution in Question 11, test the following hypotheses:

    a) $H_0: \mu_1 \geq 6$
       $H_1: \mu_1 < 6, \alpha = 0.05$
       $\bar{X} = 1.0$

    b) $H_0: \mu_1 = \mu_0 = 6$
       $H_1: \mu_1 \neq \mu_0, \alpha = 0.05$
       $\bar{X} = 1.5$

    c) $H_0: \mu_1 = \mu_0 = 6$
       $H_1: \mu_1 \neq \mu_0, \alpha = 0.05$
       $\bar{X} = 11.0$

    d) $H_0: \mu_1 \leq 6$
       $H_1: \mu_1 > 6, \alpha = 0.05$
       $\bar{X} = 10.5$

    e) $H_0: \mu_1 = \mu_0 = 6$
       $H_1: \mu_1 \neq 6, \alpha = 0.05$
       $\bar{X} = 11.5$

    f) $H_0: \mu_1 \neq \mu_0 = 6$
       $H_1: \mu_1 \neq 6, \alpha = 0.01$
       $\bar{X} = 0.5$

    g) $H_0: \mu_1 \geq 6$
       $H_1: \mu_1 < 6, \alpha = 0.01$
       $\bar{X} = 0.5$

Tests of Significance, One-Sample Case

Occasionally, a researcher will select a single sample from some un-known population and raise such questions as, "Is it reasonable to hypothesize that this sample was drawn from a population with a mean of, say, 50, or 100, or 500?" Note that, in the one-sample case, there is no control condition against which to compare the mean of an experimental group. The null hypothesis usually states that the population mean from which the sample was drawn is some specific value. In the case of the one-sample test for Pearson r, the null hypothesis usually states that the population correlation coefficient (ρ) is zero.

TESTING HYPOTHESES ABOUT THE MEAN WHEN PARAMETERS ARE KNOWN

When the population mean (μ) and standard deviation (σ) are known for a given population, the appropriate test statistic is z, defined as follows:

$$z = \frac{\bar{X} - \mu_0}{\sigma_X},$$

in which

$\bar{X}$ is the mean of the sample,

μ_0 is the population mean hypothesized under the null hypothesis,

$\sigma_{\bar{X}}$ is the standard error of the mean.

Examples

Given $\bar{X} = 110$, $\sigma_{\bar{X}} = 5$, $\mu_0 = 100$; then

$$z = \frac{110 - 100}{5} = 2.00.$$

Given $\bar{X} = 45$, $\sigma_{\bar{X}} = 2.5$, $\mu_0 = 52$; then

$$z = \frac{45 - 52}{2.5} = -2.8.$$

Given $\bar{X} = 520$, $\sigma_{\bar{X}} = 12$, $\mu_0 = 500$; then

$$z = \frac{520 - 500}{12} = 1.67.$$

Do Exercises 1 through 5 at the right.

The standard error of the mean ($\sigma_{\bar{X}}$) is defined as follows:

$$\sigma_{\bar{X}} = \frac{\sigma}{\sqrt{N}},$$

OBJECTIVES

1.   Know how to test hypotheses about the mean when the population standard deviation is known.

2.   Know how to calculate the Student t-ratio when the population standard deviation is not known.

3.   Know how to calculate the Sandler A-statistic as an alternative to Student t in the one-sample case.

4.   Know how to test hypotheses concerning the population correlation coefficient.

EXERCISES

1.   Given $\bar{X} = 70$, $\sigma_{\bar{X}} = 6$, $\mu_0 = 60$; find z.

2.   Given $\bar{X} = 20$, $\sigma_{\bar{X}} = 4.05$, $\mu_0 = 27$; find z.

3.   Given $\bar{X} = 230$, $\sigma_{\bar{X}} = 25$, $\mu_0 = 250$; find z.

4.   Given $\bar{X} = 850$, $\sigma_{\bar{X}} = 45$, $\mu_0 = 750$; find z.

5.   Given $\bar{X} = 35$, $\sigma_{\bar{X}} = 10$, $\mu_0 = 52$; find z.

3

in which

σ is the population standard deviation,
N is the sample size.

Examples

Given $\sigma = 20$, $N = 30$; then

$$\sigma_{\bar{X}} = \frac{20}{\sqrt{30}} = 3.65.$$

Given $\sigma = 20$, $N = 200$; then

$$\sigma_{\bar{X}} = \frac{20}{\sqrt{200}} = 1.41.$$

Given $\sigma = 100$, $N = 120$; then

$$\sigma_{\bar{X}} = \frac{100}{\sqrt{120}} = 9.13.$$

Do Exercises 6 through 10 at the right.

We use the standard normal curve as the sampling distribution for investigating null hypotheses when the population standard deviation is known and will have a known or hypothesized population mean.

Example

In a school district, we know the mean and standard deviation of high school juniors on a standard educational test. They are $\mu = 527$ and $\sigma = 80$. A sample of fifty "special" students yielded a mean of 550. Is it reasonable to assume that this group is representative of high school juniors within the district? Stated another way, is it reasonable that a sample mean of 550 was drawn from a population in which the "true" mean is 527?

1. *Null hypothesis* (H_0): The mean of the population (μ) from which the sample was drawn equals 527, that is, $\mu = \mu_0 = 527$.

2. *Alternative hypothesis* (H_1): The mean of the population from which the sample was drawn does not equal 527, that is, $\mu \neq \mu_0$. Note that H_1 is nondirectional, so a two-tailed test of significance will be employed.

3. *Significance level*: $\alpha = 0.01$. If the difference between the sample mean and the specified population mean is so extreme that its associated probability of occurrence under H_0 is equal to or less than 0.01, we shall reject H_0.

6. Given $\sigma = 16$, $N = 10$; find $\sigma_{\bar{X}}$.

7. Given $\sigma = 16$, $N = 100$; find $\sigma_{\bar{X}}$.

8. Given $\sigma = 16$, $N = 1000$; find $\sigma_{\bar{X}}$.

9. Given $\sigma = 100$, $N = 25$; find $\sigma_{\bar{X}}$.

10. Given $\sigma = 100$, $N = 625$; find $\sigma_{\bar{X}}$.

4. *Sampling distribution*: Since the population mean and standard deviation are known, we shall employ the standard normal curve.

5. *Critical region for rejection of H_0*: In Table A, we find that $z = \pm 2.58$ includes 1% of the area beyond it. Thus, if obtained z is equal to or greater than 2.58 or equal to or less than -2.58, we shall reject H_0.

The steps in the test of significance are as follows:

Step 1. Find

$$\sigma_{\bar{X}} = \frac{\sigma}{\sqrt{N}}.$$

In the present problem, $\sigma = 80$, $N = 50$. Thus

$$\sigma_{\bar{X}} = \frac{80}{\sqrt{50}} = \frac{80}{7.07} = 11.32.$$

Step 2. Find

$$z = \frac{\bar{X} - \mu_0}{\sigma_{\bar{X}}}.$$

From Step 1, $\sigma_{\bar{X}} = 11.32$, and therefore

$$z = \frac{550 - 527}{11.32}$$

$$= 2.03.$$

Conclusion: Since $z = 2.03$ is not within the critical region, we fail to reject H_0.

Do Exercise 11 at the right.

TESTING HYPOTHESES ABOUT THE MEAN WHEN PARAMETERS ARE UNKNOWN

Estimating the Standard Error of the Mean

When the population standard deviation is not known, it must be estimated from the sample standard deviation. The sample standard deviation is, in turn, employed to estimate the standard error of the

11. A sample of forty subjects obtained $\bar{X} = 60$ on a task in which μ and σ are known to be 70 and 15 respectively. Set this problem up in formal statistical terms and test the null hypothesis, employing $\alpha = 0.05$, two-tailed test.

3

mean. The symbol for the standard error of the mean that is estimated from the sample standard deviation is $s_{\bar{X}}$. It is defined as

$$s_{\bar{X}} = \frac{s}{\sqrt{N}},$$

in which

$$s = \sqrt{\frac{\sum x^2}{N-1}} \text{ and}$$

N = sample size.

A useful computational formula is:

$$s_{\bar{X}} = \sqrt{\frac{\sum x^2}{N(N-1)}},$$

in which

$$\sum x^2 = \sum X^2 - \frac{(\sum X)^2}{N}.$$

Example

Given the following scores, find $s_{\bar{X}}$: 18, 10, 15, 23, 17.

$$\sum X^2 = 1467$$

$$\sum X = 83$$

$$\sum x^2 = \sum X^2 - \frac{(\sum X)^2}{N}$$

$$= 1467 - \frac{(83)^2}{5}$$

$$= 1467 - 1377.8$$

$$= 89.2$$

$$s_{\bar{X}} = \sqrt{\frac{\sum x^2}{N(N-1)}}$$

$$= \sqrt{\frac{89.2}{5(4)}}$$

$$= 2.11$$

Do Exercises 12 through 15 at the right.

Find $s_{\bar{X}}$ for the following sets of scores.

12. 15, 27, 10, 19, 26, 15

13. 30, 25, 42, 49, 53, 36, 44, 57

14. 2, 8, 5, 9, 13, 11, 10, 14, 11, 7

15. 5, 19, 3, 42, 27, 1, 37, 13, 21, 32, 12

The Student *t*-Ratio

The appropriate test statistic when the population standard deviation is unknown is the Student *t*-ratio:

$$t = \frac{\bar{X} - \mu_0}{s_X},$$

in which

$\bar{X}$ is the mean of the sample,
μ_0 is the mean of the population hypothesized under H_0,
$s_{\bar{X}}$ is the estimated standard error of the mean.

Examples

Given $\bar{X} = 30$, $\mu_0 = 24$, $s_X = 3.62$; then

$$t = \frac{30 - 24}{3.62} = \frac{6}{3.62} = 1.66.$$

Given $\bar{X} = 92$, $\mu_0 = 100$, $s_{\bar{x}} = 4.03$; then

$$t = \frac{92 - 100}{4.03} = -1.99.$$

Do Exercises 16 through 19 at the right.

The *t*-Distributions

Like the standard normal curve, the *t*-distributions are symmetrical about a mean of zero. However, they are more spread out than the standard normal curve. Consequently a greater proportion of area will lie beyond a given *t* than beyond a *z* of the same value. However, as N increases, the *t*-distributions increasingly resemble the standard normal curve.

There is a separate *t*-distribution for each value of N. Each distribution is plotted in terms of degrees of freedom (df). In the one-sample case, df $= N - 1$.

Examples

If $N = 20$, then df $= 19$.
If $N = 105$, then df $= 104$.
If $N = 8$, then df $= 7$.
If $N = 5$, then df $= 4$.

Do Exercises 20 through 29 at the right.

Table B shows the critical values for rejecting H_0 at the 0.05 and

16. Given $\bar{X} = 470$, $\mu_0 = 500$, $s_{\bar{x}} = 12$; find *t*.

17. Given $\bar{X} = 55$, $\mu_0 = 50$, $s_{\bar{x}} = 2.83$; find *t*.

18. Given $\bar{X} = 107$, $\mu_0 = 100$, $s_{\bar{x}} = 3.05$; find *t*.

19. Given $\bar{X} = 80$, $\mu_0 = 88$, $s_{\bar{x}} = 3.57$; find *t*.

Specify the degrees of freedom for the following sample sizes.

20. 18

21. 2

22. 92

23. 108

24. 100

25. 11

26. 42

27. 31

28. 3

29. 72

0.01 levels, one- and two-tailed alternative hypotheses. Obtained t is significant if its absolute value (i.e., ignoring the minus sign when t is negative) equals or exceeds the tabled critical value.

Examples

Given $H_0: \mu = \mu_0, H_1: \mu \neq \mu_0, \alpha = 0.05, N = 21$, and $t = 1.99$, do we reject H_0?

In Table B, under 0.05, two-tailed test and df $= 20$, we find a critical value of 2.086. Since obtained $t = 1.99$ is less than the critical value of 2.086, we fail to reject H_0.

Given $H_0: \mu \geq \mu_0, H_1: \mu < \mu_0, \alpha = 0.05, N = 21$, and $t = -1.99$, do we reject H_0?

In Table B, under 0.05, one-tailed test and df $= 20$, we find a critical value of 1.725. Since the absolute value of obtained t is 1.99, it exceeds the critical value of 1.725, and we reject H_0.

Do Exercises 30 through 32 at the right.

A Worked Example

Consumer groups have often accused automobile manufacturers of installing odometers that overestimate mileage traveled in order to make the miles-per-gallon figure look more attractive. (For example, if a car travels 100 miles on 8 gallons of gas, the miles-per-gallon is 12.5. However, if the odometer reads, say, 109 miles, the automobile owner will erroneously conclude that his miles-per-gallon figure is 13.5.) To test this hypothesis, 12 new cars were run over a measured course of 100 miles. The odometer readings were: 103, 97, 101, 100, 99, 104, 102, 103, 100, 102, 105, 103. Test the null hypothesis, using $\alpha = 0.05$, one-tailed test.

1. *Null hypothesis* (H_0): The population mean from which the odometer samples were drawn is equal to or less than 100, that is, $\mu_0 \leq 100$.

2. *Alternative hypothesis* (H_1): The population mean is greater than 100, that is, $\mu_0 > 100$.

3. *Statistical test*: Since we are involved in a comparison of a sample mean with an hypothesized value and σ is not known, the Student t-ratio, one-sample case, is appropriate.

4. *Significance level*: $\alpha = 0.05$.

5. *Sampling distribution*: The sampling distribution is the Student t-distribution with df $= N - 1 = 12 - 1 = 11$.

6. *Critical region*: By referring to Table B, we find that the critical value for significance at 0.05 level, one-tailed test, when df $= 11$

30. Given $H_0: \mu = \mu_0, H_1: \mu \neq \mu_0$, $\alpha = 0.01, N = 31$, and $t = 2.48$; test H_0 and make the statistical decision.

31. Given $H_0: \mu \leq \mu_0, H_1: \mu > \mu_0$, $\alpha = 0.01, N = 61$, and $t = 2.44$; test H_0 and make the statistical decision.

32. Given $H_0: \mu = \mu_0, H_1: \mu \neq \mu_0$, $\alpha = 0.05, N = 12$, and $t = 2.25$; test H_0 and make the statistical decision.

is 1.796. If the value of obtained t equals or exceeds $+1.796$, we shall reject H_0 and assert H_1.

The steps in the test of significance are as follows:

Step 1. Find $\sum X$, and divide by N to obtain $\bar{X}$:

$$\sum X = 1219,$$

$$\bar{X} = 101.58.$$

Step 2. Find the sum of squares $\left(\sum x^2\right)$ by squaring each score, summing, and subtracting $\left(\sum X\right)^2 / N$:

$$\sum x^2 = \sum X^2 - \frac{\left(\sum X\right)^2}{N}$$

$$= 123{,}887 - 123{,}830.08$$

$$= 56.92.$$

Step 3. Find $s_{\bar{X}}$:

$$s_{\bar{X}} = \sqrt{\frac{\sum x^2}{N(N-1)}}$$

$$= \sqrt{\frac{56.92}{132}}$$

$$= 0.66.$$

Step 4. Find t:

$$t = \frac{\bar{X} - \mu_0}{s_{\bar{X}}}$$

$$= \frac{101.58 - 100}{0.66}$$

$$= 2.39.$$

Since obtained $t = 2.39$ exceeds the critical value of $+1.796$, we reject H_0. It would appear that the odometers are from a population with a mean greater than 100.

Do Exercise 33 at the right.

The Sandler *A*-Statistic

The Sandler A is an algebraically equivalent alternative to the Student t-ratio. The writer has shown that it can be employed in the one-sample case as a direct substitute for the Student t-ratio. The main virtue of the A-statistic is its ease of computation:

$$A = \frac{\sum D^2}{\left(\sum D\right)^2}.$$

To illustrate the calculation of A, we shall use the previous data on odometer readings.

33. Given the following scores: 8, 12, 15, 9, 13, 11, 17, 7, 11, 12, 14, 10, 12, 13. Test $H_0: \mu = \mu_0 = 10$. Employ $\alpha = 0.05$.

X	$(X - \mu_0)$	D^2
103	3	9
97	-3	9
101	1	1
100	0	0
99	-1	1
104	4	16
102	2	4
103	3	9
100	0	0
102	2	4
105	5	25
103	3	9
	$\sum D = 19$	$\sum D^2 = 87$

Step 1. Subtract μ_0 from each score, enter into the D column. Sum to obtain $\sum D = 19$.

Step 2. Square each D and place in the D^2 column. Sum to obtain $\sum D^2 = 87$.

Step 3. Find A by dividing $\sum D^2$ by $(\sum D)^2$:

$$A = \frac{87}{(19)^2} = \frac{87}{361} = 0.240.$$

Looking in Table D under 11 df and 0.05 level, one-tailed test, we find that A *equal to or less than* 0.368 falls in the critical region for rejecting H_0. Thus we reject H_0 and assert H_1.*

Do Exercise 34 at the right.

---

* The probability values of both t and A are identical since A is derived from t. Thus

$$t = \sqrt{\frac{(N-1)}{NA - 1}}.$$

In the present problem,

$$t = \sqrt{\frac{11}{2.88 - 1}} = \sqrt{\frac{11}{1.88}} = \sqrt{5.85} = 2.42.$$

This agrees with our previously calculated $t = 2.39$, allowing for a rounding error of 0.03.

34. Calculate Sandler A for the data in Exercise 35, testing the same H_0 at $\alpha = 0.05$.

TEST OF SIGNIFICANCE FOR PEARSON r, ONE-SAMPLE CASE

When a correlation coefficient is calculated from a bivariate distribution, it may be regarded as an estimate of the population correlation coefficient (ρ). As with all sample statistics, correlation coefficients will distribute themselves about the population value. Researchers often raise questions such as, "Is it reasonable to assume that the sample correlation coefficient was drawn from a bivariate population in which the true correlation is, say, 0.08, 0.20, -0.65, etc.?" The most common null hypothesis investigated is that the sample was drawn from a population in which the correlation is zero.

The test statistic is the normal deviate, z,

$$z = \frac{z_r - \mathcal{Z}_r}{\sqrt{\dfrac{1}{n-3}}},$$

where

z_r = the transformed value of the sample r,
$\mathcal{Z}_r$ = the transformed value of the population correlation coefficient specified under H_0,
n = the number of paired scores.

The transformed values of z_r and $\mathcal{Z}_r$ are found in Table F.

Examples

Given $r = 0.56$, $H_0: \rho = 0.00$, $n = 20$; then

$$z = \frac{0.633 - 0.00}{\sqrt{\dfrac{1}{17}}}$$

$$= \frac{0.633}{0.243}$$

$$= 2.60.$$

Given $r = 0.62$, $H_0: \rho = 0.50$, $n = 125$; then

$$z = \frac{0.725 - 0.549}{\sqrt{\dfrac{1}{122}}}$$

$$= \frac{0.176}{0.091}$$

$$= 1.93.$$

3

Do Exercises 35 through 38 at the right.

The standard normal curve is used to evaluate H_0.

Example

Given $r = 0.56$, $H_0: \rho = 0.00$, $n = 20$, $\alpha = 0.01$, two-tailed test, $z = 2.60$.

Since $z = 2.60$ is in the critical region at $\alpha = 0.01$, we reject H_0 and assert that the sample was drawn from a bivariate distribution with a population correlation coefficient greater than zero.

Do Exercises 39 through 42 at the right.

35. Given $r = 0.45$, $H_0: \rho = 0.00$, $n = 26$; find z.

36. Given $r = -0.82$, $H_0: \rho = 0.00$, $n = 10$; find z.

37. Given $r = 0.15$, $H_0: \rho = 0.00$, $n = 500$; find z.

38. Given $r = 0.65$, $H_0: \rho = 0.60$, $n = 100$; find z.

39. Refer to Exercise 35. Using $\alpha = 0.05$, one-tailed test, make the statistical decision.

40. Refer to Exercise 36. Using $\alpha = 0.01$, two-tailed test, make the statistical decision.

41. Refer to Exercise 37. Using $\alpha = 0.05$, two-tailed test, make the statistical decision.

42. Refer to Exercise 38. Using $\alpha = 0.05$, one-tailed test, make the statistical decision.

CHAPTER 3 TEST

To complete this test, students will require access to Tables A (Percent of Areas under the Standard Normal Curve), B (Critical Values of t), and F (Transformation of r' to z_r).

1. Given $\bar{X} = 200$, $\sigma_{\bar{X}} = 12$, and $\mu_0 = 220$; z equals:

   a) -5.78 b) -1.67 c) 5.78 d) 0.60

2. Given $\bar{X} = 100$, $\sigma_{\bar{X}} = 16$, and $\mu_0 = 90$; z equals:

   a) 0.62 b) 1.60 c) -1.60 d) 2.50

3. Given $\sigma = 40$, $N = 64$; $\sigma_{\bar{X}}$ equals:

   a) 0.62 b) 0.79 c) 6.33 d) 5.00

4. Given $\sigma = 12$; $N = 49$; $\sigma_{\bar{X}}$ equals:

   a) 1.71 b) 4.08 c) 0.58 d) 14.16

5. Given $\sigma = 80$, $N = 625$, $\mu_0 = 350$, $\bar{X} = 356$; z equals:

   a) 16.67 b) -1.88 c) 1.88 d) 4.38

6. Given $\sigma = 36$, $N = 81$, $\mu_0 = 148$, $\bar{X} = 140$; z equals:

   a) -2.00 b) 18.18 c) 2.00 d) -1.82

7. Given $H_0: \mu_1 = \mu_0$, $H_1: \mu_1 \neq \mu_0$, $\alpha = 0.05$, and we reject H_0; the absolute value of the z-statistic must have equalled or been beyond what value?

   a) 1.96 b) 1.65 c) 2.58 d) 2.33

8. Given $\mu_0 = 130$, $\mu_1 = 150$, $s = 25$, and $N = 45$; what test statistic is appropriate?

   a) z

   b) t

   c) z_r

   d) none of the preceding

9. Given the following set of scores: 0, 2, 4, 6; $s_{\bar{X}}$ equals:

   a) 20 b) 4.67 c) 2.16 d) 1.29

10. Given $\sum x^2 = 145$, $N = 9$; $s_{\bar{X}}$ equals:

    a) 16.11 b) 4.01 c) 1.42 d) 2.01

11. Given $\sum x^2 = 16$, $N = 6$; $s_{\bar{X}}$ equals:

    a) 0.73 b) 5.37 c) 2.32 d) 0.53

12. Given $\bar{X} = 120$, $\mu_0 = 100$, $s_{\bar{X}} = 6.95$, $N = 25$; t equals:

    a) -14.39 b) 2.88 c) 14.39 d) 0.35

3

13. Given $H_0: \mu_1 = \mu_0$, $H_1: \mu_1 \neq \mu_0$, $\alpha = 0.05$, $t = -2.08$, $N = 26$; our statistical decision is to:

   a) reject H_0 and assert H_1

   b) reject H_0 but not assert H_1

   c) fail to reject H_0 but assert H_1

   d) fail to reject H_0 and, therefore, not assert H_1

14. Given $\sum D = 60$, $\sum D^2 = 205$; A equals:

   a) 0.2927   b) 3.42   c) 0.0569   d) 17.56

15. Given $r = -0.53$, $H_0: \rho \geq 0.00$, $H_1: \rho < 0.00$, $N = 52$; z equals:

   a) 0.08   b) -4.21 c) -0.08 d) 4.21

Practical

16. Given $\bar{X} = 53$, $\mu_0 = 50$, $\sigma = 9$, and $N = 20$; test $H_0: \mu_1 = \mu_0$ at $\alpha = 0.01$.

17. Given $\bar{X} = 53$, $\mu_0 = 50$, $\sum X = 742$, $\sum X^2 = 39{,}396$, and $N = 14$; test $H_0: \mu_1 \leq 50$, using $\alpha = 0.01$.

18. Given $\mu_0 = 30$ and the data that follow; test $H_0: \mu = \mu_0$ at $\alpha = 0.01$.

X	X
35	31
33	29
38	32
34	27
32	34
30	36

19. Given $r = 0.55$, $H_0: \rho \leq 0.20$, $N = 60$; test H_0 using $\alpha = 0.01$.

Significance of Difference Between Means, Independent Samples

When subjects or observations are drawn at random from a given population and are assigned at random to two groups, we speak of such a research design as the two-sample case with independent samples. Typically, one group (the experimental group) receives the experimental treatment and the other group (the control group) does not receive the experimental treatment. In drug research, the control group is commonly administered a placebo, which is usually an inert ingredient that resembles the experimental drug in appearance. Scores are obtained on some criterion measure (e.g., performance on a psychomotor task, or change in a diagnosed medical condition); then the means of the two groups are calculated, and a test of significance is applied to determine if both groups might reasonably be considered representative of the same population.

TEST OF SIGNIFICANCE WHEN PARAMETERS ARE KNOWN

It is not often that we know the mean, standard deviation, and variance of a population. When they are known, the appropriate test of significance is

$$z = \frac{(\bar{X}_1 - \bar{X}_2) - (\mu_1 - \mu_2)}{\sigma_{\bar{X}_1 - \bar{X}_2}},$$

in which

$\bar{X}_1 - \bar{X}_2$ is the difference in sample means,
$\mu_1 - \mu_2$ is the difference in the population means from which the samples were presumed to have been drawn,
$\sigma_{\bar{X}_1 - \bar{X}_2}$ is the standard error of the difference between means when the population variance is known.

Examples

To test the hypothesis that μ_1 is ten points higher than μ_2, the z-statistic becomes

$$z = \frac{(\bar{X}_1 - \bar{X}_2) - 10}{\sigma_{\bar{X}_1 - \bar{X}_2}}.$$

To test the hypothesis that μ_1 is fifteen points lower than μ_2, the z-statistic becomes

$$z = \frac{(\bar{X}_1 - \bar{X}_2) - (-15)}{\sigma_{\bar{X}_1 - \bar{X}_2}}$$

$$= \frac{(\bar{X}_1 - \bar{X}_2) + 15}{\sigma_{\bar{X}_1 - \bar{X}_2}}.$$

Do Exercises 1 and 2 at the right.

OBJECTIVES

1. Know when and how to use z as the test statistic for ascertaining the significance of the difference between means.

2. Know when and how to use t as the test statistic for ascertaining the significance of the difference between means.

3. Know how to apply the test for homogeneity of variances.

4. Know how to estimate the degree of association between the experimental and dependent variables.

EXERCISES

In the following exercises, assume that the population mean, standard deviation, and variance are known.

1. Set up the test statistic to investigate the null hypothesis that the sample means were drawn from two populations in which μ_2 is 6.5 points higher than μ_1.

2. Set up the test statistic to investigate the hypothesis that the sample means were drawn from populations in which μ_1 is 9 points higher than μ_2.

4

The most common null hypothesis is that the samples were drawn from the same population or from populations with identical means. In other words, H_0 becomes $\mu_1 = \mu_2$ or, alternatively, $\mu_1 - \mu_2 = 0$. To investigate the null hypothesis of no difference between population means, the z-statistic simplifies to

$$z = \frac{\bar{X}_1 - \bar{X}_2}{\sigma_{\bar{X}_1 - \bar{X}_2}}.$$

Do Exercise 3 at the right.

The Standard Error of the Difference ($\sigma_{\bar{X}_1 - \bar{X}_2}$)

When parameters are known, the standard error of the difference between means ($\sigma_{\bar{X}_1 - \bar{X}_2}$) is

$$\sigma_{\bar{X}_1 - \bar{X}_2} = \sqrt{\frac{\sigma_1^2}{N_1} + \frac{\sigma_2^2}{N_2}},$$

in which σ_1^2 and σ_2^2 are the variances of the populations and N_1 and N_2 are the sample sizes drawn from their respective populations.

Examples

Given $\sigma_1^2 = 144$, $N_1 = 25$, $\sigma_2^2 = 64$, $N_2 = 30$; then

$$\begin{aligned}
\sigma_{\bar{X}_1 - \bar{X}_2} &= \sqrt{\frac{144}{25} + \frac{64}{30}} \\
&= \sqrt{5.76 + 2.13} \\
&= \sqrt{7.89} \\
&= 2.81.
\end{aligned}$$

Given $\sigma_1 = 18$, $N_1 = 35$, $\sigma_2 = 26$, $N_2 = 35$; then

$$\begin{aligned}
\sigma_{\bar{X}_1 - \bar{X}_2} &= \sqrt{\frac{324}{35} + \frac{676}{35}} \\
&= \sqrt{\frac{1000}{35}} \\
&= \sqrt{28.57} \\
&= 5.35.
\end{aligned}$$

Do Exercises 4 and 5 at the right.

Testing the Null Hypothesis

The standard normal curve is the appropriate sampling distribution

3. Set up the test statistic to investigate the null hypothesis that the sample means, $\bar{X}_A$ and $\bar{X}_B$, were drawn from the same population of means.

4. Given

$$\sigma_1^2 = 400, N_1 = 15, \sigma_2^2 = 290,$$
$$N_2 = 15;$$

find $\sigma_{\bar{X}_1 - \bar{X}_2}$.

5. Given

$$\sigma_1 = 32, N_1 = 17, \sigma_2 = 53, N_2 = 26;$$

find $\sigma_{\bar{X}_1 - \bar{X}_2}$.

for testing the null hypothesis of no difference between means when parameters are known.

Example

Given $\bar{X}_1 = 25.08$, $\bar{X}_2 = 17.12$, $\sigma_1 = 12$, $\sigma_2 = 8$, $N_1 = 25$, and $N_2 = 30$. Use a two-tailed or nondirectional alternative hypothesis.

Step 1. Specify the null hypothesis. In the present example, H_0 is that there is no difference in the population means from which the two samples were drawn, i.e., $\mu_1 = \mu_2$.

Step 2. Specify the alternative hypothesis. Since the null hypothesis is nondirectional, the alternative hypothesis is that the population means from which the samples were drawn are not the same, i.e., $\mu_1 \neq \mu_2$.

Do Exercise 6 at the right.

Step 3. Specify the statistical test. In the present example, the z-statistic is appropriate since the population standard deviations are known.

Step 4. Specify the significance level. We shall use $\alpha = 0.05$.

Step 5. Specify the sampling distribution. In this example, the sampling distribution is the standard normal curve (Table A).

Step 6. Specify the critical region. Since the alternative hypothesis is two-tailed, we shall look for both a positive and a negative z in which 2.50 percent of the area is beyond that z. In the present example, $z \geq 1.96$ or $z \leq -1.96$ will lead to rejection of H_0.

Do Exercise 7 at the right.

Step 7. Find z. In the present example,

$$z = \frac{25.08 - 17.12}{\sqrt{\dfrac{144}{25} + \dfrac{64}{30}}}$$

$$= \frac{7.96}{2.81}$$

$$= 2.83.$$

Since $z = 2.83 > 1.96$, we reject H_0 and assert that sample 1 was drawn from a population with a higher mean than the population mean represented by sample 2.

6. Given

$$\bar{X}_1 = 6.57, \bar{X}_2 = 5.83, \sigma_1 = 8,$$
$$\sigma_2 = 11, N_1 = 15, \text{ and } N_2 = 18.$$

Specify the null and alternative hypotheses, using a two-tailed test and $\alpha = 0.05$.

7. Complete Steps 3–6 using the information found in Exercise 6.

4

Do Exercise 8 at the right.

TEST OF SIGNIFICANCE WHEN PARAMETERS ARE UNKNOWN

As previously indicated, it is extremely rare that we have knowledge of parameters. Typically, we have samples drawn from various populations and we must estimate parameters from the sample statistics themselves. When parameters are unknown, the appropriate test statistic is Student's t-ratio,

$$t = \frac{\bar{X}_1 - \bar{X}_2 - (\mu_1 - \mu_2)}{s_{\bar{X}_1 - \bar{X}_2}},$$

in which

$\bar{X}_1 - \bar{X}_2$ is the difference in sample means,

$\mu_1 - \mu_2$ is the difference in the population means from which the samples were presumed to have been drawn,

$s_{\bar{X}_1 - \bar{X}_2}$ is the standard error of the difference between means when estimated from sample variances.

Example

To test the hypothesis that μ_1 is 8 points higher than μ_2, the t-statistic becomes:

$$t = \frac{\bar{X}_1 - \bar{X}_2 - 8}{s_{\bar{X}_1 - \bar{X}_2}}.$$

Do Exercises 9 and 10 at the right.

The most common null hypothesis is that the samples were selected from the same population or from populations with identical means. In other words, H_0 becomes $\mu_1 = \mu_2$, or alternatively, $\mu_1 - \mu_2 = 0$. To investigate the null hypothesis of no difference between population means, the t-statistic simplifies to

$$t = \frac{\bar{X}_1 - \bar{X}_2}{s_{\bar{X}_1 - \bar{X}_2}}.$$

Do Exercise 11 at the right.

The Standard Error of the Difference $(s_{\bar{X}_1 - \bar{X}_2})$

When parameters are not known, the standard error of the difference between means must be estimated from the sample standard deviation or variances.

---

8. Using the information in Exercises 6 and 7, find z and draw the appropriate conclusion.

9. Show the t-statistic for testing the hypothesis that μ_1 is 12 points higher than μ_2.

10. Show the t-statistic for testing the hypothesis that μ_2 is 7 points higher than μ_1.

11. Set up the test statistic to investigate the null hypothesis that the sample means $\bar{X}_A$ and $\bar{X}_B$ were drawn from the same population of means.

When n's are equal.   When the n in each group is the same, the convenient computational formula for $s_{\bar{X}_1 - \bar{X}_2}$ is

$$s_{\bar{X}_1 - \bar{X}_2} = \sqrt{\frac{\sum x_1^2 + \sum x_2^2}{n(n-1)}},$$

in which

$\sum x_1^2$ is the sum of squares of condition X_1 and is defined as

$$\sum x_1^2 = \sum X_1^2 - \frac{(\sum X_1)^2}{n_1},$$

$\sum x_2^2$ is the sum of squares of condition X_2 and is defined as

$$\sum x_2^2 = \sum X_2^2 - \frac{(\sum X_2)^2}{n_2}.$$

Examples

Given $\sum X_1 = 451, \sum X_1^2 = 20969, n_1 = 10, n_2 = 10, \sum X_2 = 491,$ $\sum X_2^2 = 24{,}953$; then

$$\sum x_1^2 = 20{,}969 - \frac{(451)^2}{10}$$

$$= 628.9;$$

$$\sum x_2^2 = 24{,}953 - \frac{(491)^2}{10}$$

$$= 844.9;$$

$$s_{\bar{X}_1 - \bar{X}_2} = \sqrt{\frac{628.9 + 844.9}{90}}$$

$$= 4.05.$$

Do Exercises 12 and 13 at the right.

When n's are unequal.   When the n in each group is not the same, a convenient formula for $s_{\bar{X}_1 - \bar{X}_2}$ is

$$s_{\bar{X}_1 - \bar{X}_2} = \sqrt{\left(\frac{\sum x_1^2 + \sum x_2^2}{n_1 + n_2 - 2}\right)\left(\frac{1}{n_1} + \frac{1}{n_2}\right)}.$$

Example

Given: $\sum X_1 = 38, \sum X_1^2 = 228, n_1 = 11, \sum X_2 = 40, \sum X_2^2 = 222,$

12. Given

$$\sum X_1 = 80, \sum X_1^2 = 904, n_1 = 9,$$
$$\sum X_2 = 90, \sum X_2^2 = 1040, n_2 = 9;$$

find $s_{\bar{X}_1 - \bar{X}_2}$.

13. Given

$$\sum X_1 = 22, \sum X_1^2 = 186, n_1 = 12,$$
$$\sum X_2 = 12, \sum X_2^2 = 106, n_2 = 12;$$

find $s_{\bar{X}_1 - \bar{X}_2}$.

4

$n_2 = 9$; then

$$\sum x_1^2 = 228 - \frac{(38)^2}{11}$$

$$= 96.73;$$

$$\sum x_2^2 = 222 - \frac{(40)^2}{9}$$

$$= 44.22;$$

$$s_{\bar{X}_1 - \bar{X}_2} = \sqrt{\frac{96.73 + 44.22}{18}\left(\frac{1}{11} + \frac{1}{9}\right)}$$

$$= \sqrt{\frac{140.95}{18}(0.20)} = \sqrt{\frac{28.19}{18}}$$

$$= 1.25.$$

Do Exercises 14 and 15 at the right.

Testing the Null Hypothesis

The Student t-distributions are the appropriate sampling distributions for testing the null hypothesis of no difference between means when parameters are not known.

Example

The manager of a school cafeteria believed that women (Group X_2) consume more coffee daily than men (Group X_1). He asked a total of 20 randomly selected students on randomly selected days how many cups of coffee they had consumed during the previous 24-hour period. He obtained the following results:

$$\bar{X}_1 = 3.45, \quad \sum X_1 = 38, \quad \sum X_1^2 = 228, \quad n_1 = 11,$$

$$\bar{X}_2 = 4.44, \quad \sum X_2 = 40, \quad \sum X_2^2 = 222, \quad n_2 = 9.$$

Use $\alpha = 0.05$, two-tailed test.

Step 1. Specify the null hypothesis. In the present example, H_0 is that there is no difference in the population means of daily cups of coffee consumed by male and female students, i.e., $\mu_1 = \mu_2$.

Step 2. Specify the alternative hypothesis. Since the null hypothesis is nondirectional, the alternative hypothesis is that the population means from which samples were drawn are not the same, i.e., $\mu_1 \neq \mu_2$.

14. Given

$$\sum X_1 = 33, \sum X_1^2 = 149, n_1 = 10,$$
$$\sum X_2 = 20, \sum X_2^2 = 80, n_2 = 8;$$

find $s_{\bar{X}_1 - \bar{X}_2}$.

15. Given

$$\sum X_1 = 120, \sum X_1^2 = 1472, n_1 = 12,$$
$$\sum X_2 = 135, \sum X_2^2 = 2016, n_2 = 10;$$

find $s_{\bar{X}_1 - \bar{X}_2}$.

Do Exercise 16 at the right.

Step 3. Specify the statistical test. In the present example, the *t*-statistic is appropriate since the population standard deviations are not known.

Step 4. Specify the significance level. We shall use $\alpha = 0.05$.

Step 5. Specify the sampling distribution. In the present example, the sampling distribution is the Student *t*-distribution with df $= n_1 + n_2 - 2$ or $11 + 9 - 2 = 18$ (Table B).

Step 6. Specify the critical region. Since H_1 is nondirectional, the critical region consists of all values of $t \geq 2.101$ and $t \leq -2.101$.

Do Exercise 17 at the right.

Step 7. Find *t*. In the present example;

$$t = \frac{3.45 - 4.44}{\sqrt{\frac{\sum x_1^2 + \sum x_2^2}{n_1 + n_2 - 2}\left(\frac{1}{n_1} + \frac{1}{n_2}\right)}}$$

and

$$\sum x_1^2 = 228 - \frac{(38)^2}{11}$$

$$= 96.73,$$

$$\sum x_2^2 = 222 - \frac{(40)^2}{9}$$

$$= 44.22.$$

Therefore

$$s_{\bar{x}_1 - \bar{x}_2} = \sqrt{\frac{96.73 + 44.22}{18}\left(\frac{1}{11} + \frac{1}{9}\right)}$$

$$= 1.25,$$

and the *t*-ratio becomes

$$t = -\frac{0.99}{1.25}$$

$$= -0.792.$$

Since $t = -0.792$ is not within the critical region, we fail to reject H_0. The results of this study do not permit us to conclude that women students drink more coffee daily than men students.

16. A group of subjects agreed to participate in a study to determine the effects of a drug on a task of psychomotor coordination. The results are shown for the drug group (X_1) and for the placebo group (X_2):

$\bar{X}_1 = 26.91, \sum X_1 = 296, \sum X_1^2 = 8209,$

$n_1 = 11, \bar{X}_2 = 24.62, \sum X_2 = 320,$

$\sum X_2^2 = 8361, n_2 = 13.$

Specify the null and alternative hypotheses, using a two-tailed test and $\alpha = 0.01$.

17. Complete Steps 3–6, using the information found in Exercise 16.

4

Do Exercise 18 at the right.

18. Find t and draw the appropriate conclusion.

A Worked Example

Nondirectional alternative hypothesis. An investigator believed that the life of batteries is affected by the way they are used. One sample was subjected to continuous use; the other, to frequent rest periods. He tested both samples of batteries with the same kind of light bulbs and recorded the total number of life hours. What did he conclude?

Continuous		Rest periods	
20	18	20	20
21	19	20	26
19	20	25	22
17	18	23	24
20	19	24	25

1. *Null hypothesis* (H_0): There is no difference between the population means of the continuous use versus the use of batteries with rest periods, i.e., $\mu_1 = \mu_2$.

2. *Alternative hypothesis* (H_1): There is a difference in life expectancies of batteries, depending on their use. Since the alternative hypothesis is nondirectional, a two-tailed test of significance is used.

3. *Statistical test*: Since we are comparing two sample means presumably drawn from normally distributed populations with equal variances, the Student t-ratio, two-sample case, is appropriate.

4. *Significance level*: $\alpha = 0.01$.

5. *Sampling distribution*: t-distribution with df $= n_1 + n_2 - 2$ or 18.

6. *Critical region*: Since H_1 is nondirectional, we consult Table B, 0.01 significance level, two-tailed test, at 18 df. We find $t \geq 2.878$ or $t \leq -2.878$ is required to reject H_0.

7. Since $n_1 = n_2$, the following formula for $s_{\bar{X}_1 - \bar{X}_2}$ is used:

$$s_{\bar{X}_1 - \bar{X}_2} = \sqrt{\frac{\sum x_1^2 + \sum x_2^2}{n(n-1)}}.$$

Since

$$\sum x_1^2 = 3661 - \frac{(191)^2}{10}$$

$$= 12.9,$$

$$\sum x_2^2 = 5291 - \frac{(229)^2}{10}$$

$$= 46.9,$$

we obtain

$$s_{\bar{X}_1 - \bar{X}_2} = \sqrt{\frac{12.9 + 46.9}{10(9)}}$$

$$= \sqrt{0.6644}$$

$$= 0.82$$

and therefore

$$t = \frac{19.1 - 22.9}{0.82}$$

$$= -4.634.$$

Since $t = -4.634$ is less than $t = -2.878$, it falls within the critical region. We can reject the null hypothesis and assert that batteries given rest periods have longer life expectancies than those run continuously.

Do Exercise 19 at the right.

Directional alternative hypothesis.   Let us imagine that, in the preceding example, the investigator had used a theory of recovery of electrical potential during periods of rest to predict, in advance of the study, that batteries given rest periods would last longer. In other words, his alternative hypothesis would have been directional: $\mu_2 > \mu_1$.

The procedures for calculating t are precisely the same. However, the formulations of Step 1 (H_0), Step 2 (H_1), and Step 6 (critical region) must be altered to reflect the directional hypothesis.

1. *Null hypothesis* (H_0): The population mean life expectancy of the continuously used batteries is equal to or greater than the life expectancy of the batteries given rest periods, i.e., $H_0: \mu_1 \geq \mu_2$.

2. *Alternative hypothesis* (H_1): The population mean life expectancy of the batteries given rest periods is greater than the life expectancy of the batteries run continuously, i.e., $H_1: \mu_2 > \mu_1$.

6. *Critical region*: Since H_1 is directional, we consult Table B, 0.01 significance level, one-tailed test, at 18 df. We find $t \leq -2.552$ is required to reject H_0.

Do Exercise 20 at the right.

19. An investigator was interested in whether people who vacation at a certain resort on the American plan (meals included) tend to gain more weight on their vacations than people at the same resort on the European plan (no meals included). He selected two random samples and recorded the change in weight over a two-week period. What did he conclude? Employ $\alpha = 0.05$.

American plan		European plan	
$+6$	0	-1	$+3$
$+9$	$+5$	0	-5
0	-3	$+4$	$+4$
-2	$+5$	0	$+1$
$+2$	-1	$+1$	-1
$+1$	0	0	$+6$

20. Rephrase Steps 1, 2, and 6 to test the alternative hypothesis that tourists on the American plan will gain more weight. Test this hypothesis and draw the appropriate conclusion.

4

THE TEST FOR HOMOGENEITY OF VARIANCE

One of the assumptions underlying the use of the t-ratio is that both samples are drawn from populations whose variances are equal. This assumption is referred to as *homogeneity of variance*. Statisticians differ in the degree to which they feel that the finding of homogeneity of variance is necessary to give validity to the use of the t-ratio. However, one point is hardly questioned. Unless variances are found to be markedly heterogeneous, the conclusions drawn from the use of the Student t-ratio are not likely to be adversely affected.

The test statistic used to test the null hypothesis of homogeneity of variances is the F-ratio:

$$F = \frac{s^2 \text{ (larger variance)}}{s^2 \text{ (smaller variance)}}.$$

Examples

If $s_1^2 = 128.96$ and $s_2^2 = 205.73$, then

$$F = \frac{205.73}{128.96} = 1.60.$$

If $s_1^2 = 7.93$ and $s_2^2 = 3.54$, then

$$F = \frac{7.93}{3.54} = 2.24$$

Do Exercises 21 through 24 at the right.

Table C is used to evaluate the significance of the F-ratio. To employ this table we should first locate the column indicating degrees of freedom (df) of the condition with the larger variance and then move downward until we locate the row corresponding to degrees of freedom of the condition with the smaller variance. The value given is the critical value at the 0.05 significance level for rejecting the null hypothesis of homogeneity of variance. *If the obtained F equals or exceeds this value*, we reject the assumption of homogeneity and assert that the sample variances do not come from a common population of variances.

Examples

a) $F = 2.96$, df 8/12

The F-ratio required for significance is 3.51. Since obtained $F < 3.51$, we fail to reject H_0 of homogeneity of variances.

b) $F = 8.69$, df 15/20.

Given the following variance estimates, calculate the F-ratio for testing homogeneity of variances.

21. $s_1^2 = 92, s_2^2 = 198$

22. $s_1^2 = 75, s_2^2 = 25$

23. $s_1^2 = 0.96, s_2^2 = 3.52$

24. $s_1^2 = 12.59, s_2^2 = 60.38$

The F-ratio required for significance is 2.57. Since obtained $F > 2.57$, we reject H_0 and assert that the variances are heterogeneous, i.e., they come from different populations of variances.

Do Exercises 25 through 28 at the right.

The number of degrees of freedom for each group is equal to $n - 1$.

Examples

If $n_2 = 21$, then df $= 20$.
If $n_1 = 16$, then df $= 15$.

Do Exercise 29 and 30 at the right.

A Worked Example

In a previous example, we found $t = -2.878$, $\sum x_1^2 = 12.9$, $n_1 = 10$, $\sum x_2^2 = 46.9$, $n_2 = 10$. In that case

$$s_1^2 = \frac{\sum x_1^2}{n_1 - 1} = \frac{12.9}{9} = 1.43, \qquad \text{df} = 9;$$

$$s_2^2 = \frac{\sum x_2^2}{n_2 - 1} = \frac{46.9}{9} = 5.21, \qquad \text{df} = 9;$$

$$F = \frac{5.21}{1.43} = 3.64, \qquad \text{df} = 9/9.$$

Reference to Table C under 9 and 9 df shows that an F-ratio equal to or greater than 4.03 is required for significance. Since the obtained $F = 3.64$ is less than 4.03, we fail to reject H_0.

Do Exercise 31 at the right.

ESTIMATING DEGREE OF ASSOCIATION BETWEEN THE EXPERIMENTAL AND DEPENDENT VARIABLES

The finding of a statistically significant t-ratio means that some degree of association exists between the experimental and dependent variables. However, the fact of statistical significance does not automatically confer "importance" to a finding. Given a sufficiently large N, even a trivial difference may be found to be statistically significant.

25. Given $F = 1.89$, df 6/15, test H_0 of homogeneity of variance. Use $\alpha = 0.05$.

26. Given $F = 6.73$, df 10/15, test H_0 of homogeneity of variance. Use $\alpha = 0.05$.

27. Given $F = 3.12$, df 9/15, test H_0 of homogeneity of variance. Use $\alpha = 0.05$.

28. Given $F = 5.02$, df 4/8, test H_0 of homogeneity of variance. Use $\alpha = 0.05$.

29. Given $n_1 = 20$, find df.

30. Given $n_2 = 13$, find df.

31. In a previous example, we found $t = 0.792$, $\sum x_1^2 = 96.73$, $n_1 = 11$, $\sum x_2^2 = 44.22$, $n_2 = 9$. Test the null hypothesis that both sample variances were drawn from a common population of variances. Use $\alpha = 0.05$.

4

One way of clarifying the "significance" of a statistically significant difference is to ascertain the extent to which variations in the experimental variable (the treatment administered) account for variations in the dependent measure. In general, the higher the degree of relationship, the greater the importance of the finding.

A measure of association is ω^2 (omega squared), which is estimated by

$$\text{est } \omega^2* = \frac{t^2 - 1}{t^2 + n_1 + n_2 - 1}.$$

Example

In a previous example, we found $t = -4.634$ with $n_1 = n_2 = 10$, so

$$\text{est } \omega^2 = \frac{21.47 - 1}{21.47 + 19}$$

$$= \frac{20.47}{40.47}$$

$$= 0.51.$$

This may be interpreted to mean that approximately 51% of the variance in the dependent measure is accounted for in terms of variations of the treatment variable. This is a high degree of association.

Do Exercises 32 through 35 at the right.

32. Find est ω^2 for Exercise 18.

33. Find est ω^2 for Exercise 19.

34. Given $t = 2.65$ with $n_1 = 8$ and $n_2 = 7$; find ω^2.

35. Given $t = 2.65$ with $n_1 = 25$ and $n_2 = 23$; find ω^2.

* If $|t|$ is less than 1.00, ω^2 will be negative. Since a negative ω^2 is meaningless, we arbitrarily set $\omega^2 = 0$ when $|t| \leq 1.00$.

CHAPTER 4 TEST

To complete this test, students require access to Table A (Percent of Areas under the Standard Normal Curve), Table B (Critical Values of t), and Table C_1 (Values of F Exceeded by 0.025 of the Values in the Sampling Distribution).

1. Given $\sigma_1^2 = 150$, $n_1 = 30$, $\sigma_2^2 = 180$, $n_2 = 30$; $\sigma_{\bar{X}_1 - \bar{X}_2}$ equals:

   a) 88   b) 2.35   c) 0.38   d) 3.32

2. Given $\sigma_1 = 20$, $n_1 = 40$, $\sigma_2 = 20$, $n_2 = 38$; $\sigma_{\bar{X}_1 - \bar{X}_2}$ equals:

   a) 4.53   b) 1.01   c) 20.53   d) 3.20

3. Given $\bar{X}_1 = 6.53$, $\bar{X}_2 = 4.44$, $\sigma_{\bar{X}_1 - \bar{X}_2} = 0.78$; z equals:

   a) 1.64   b) 2.68   c) 2.38   d) 0.37

4. Given $\bar{X}_1 = 26$, $\bar{X}_2 = 18$, $\sigma_{\bar{X}_1 - \bar{X}_2} = 3.41$, $H_0 : \mu_1 \leq \mu_2$; find z and make the statistical decision:

   a) $z = 2.35$, reject H_0

   b) $z = 4.32$, reject H_0

   c) $z = 4.32$, fail to reject H_0

   d) $z = 2.35$, fail to reject H_0

5. Given $\sum x_1^2 = 72$, $\sum x_2^2 = 53$, $n_1 = 15$, $n_2 = 15$; $s_{\bar{X}_1 - \bar{X}_2}$ equals:

   a) 8.33   b) 0.60   c) 0.77   d) 2.89

6. Given $\sum x_1^2 = 484$, $\sum x_2^2 = 263$, $n_1 = 19$, $n_2 = 14$; $s_{\bar{X}_1 - \bar{X}_2}$ equals:

   a) 92.63   b) 1.68   c) 2.99   d) 1.73

7. Given $\bar{X}_1 = 84$, $\bar{X}_2 = 77$, $n_1 = 31$, $n_2 = 41$, $s_{\bar{X}_1 - \bar{X}_2} = 3.07$, t equals:

   a) 140.00   b) 4.00   c) 2.28   d) 0.44

8. Given $H_0 : \mu_1 = \mu_2$, $H_1 : \mu_1 \neq \mu_2$, $\bar{X}_1 = 15$, $\bar{X}_2 = 23$, $s_{\bar{X}_1 - \bar{X}_2} = 3.56$, $n_1 = 19$, $n_2 = 23$, and $\alpha = 0.05$; find t and make the statistical decision:

   a) $t = -4.23$, reject H_0

   b) $t = -2.25$, fail to reject H_0

   c) $t = 4.23$, reject H_0

   d) $t = -2.25$, reject H_0

9. Given $s_1^2 = 36$, $s_2^2 = 142$, $n_1 = 16$, $n_2 = 13$, $\alpha = 0.05$; find F and make the statistical decision:

   a) $F = 3.94$, fail to reject H_0

b) $F = 4.55$, reject H_0

c) $F = 4.55$, fail to reject H_0

d) $F = 3.94$, reject H_0

10.   Given $t = 1.84$, $n_1 = 6$, $n_2 = 7$; ω^2 equals:

     a)   0.22   b)   0.21   c)   0.16   d)   0.13

Practical

11.   Given $\sigma_1^2 = 108$, $\sigma_2^2 = 110$, $\bar{X}_1 = 38.46$, $\bar{X}_2 = 35.42$, $n_1 = 30$, $n_2 = 33$; test $H_0 : \mu_1 \neq \mu_2$, using $\alpha = 0.01$.

12.   The following scores were obtained by two groups of subjects. Group X_1 received the experimental treatment whereas Group X_2 served as the control condition. Test $H_0 : \mu_1 \leq \mu_2$, using $\alpha = 0.01$.

X_1	X_2
7	6
15	3
12	7
19	15
25	9
14	4
10	5
16	2
9	

13.   Test the hypothesis of homogeneity of variances for the data in Question 12.

14.   Given the following values of t, n_1, and n_2, find ω^2.

     a) $t = 2.63$, $n_1 = 5$, $n_2 = 4$

     b) $t = 2.63$, $n_1 = 10$, $n_2 = 8$

     c) $t = 2.63$, $n_1 = 15$, $n_2 = 12$

Significance of Difference Between Pearson r's, Independent Samples

There are times when we obtain two independent correlation coefficients between two variables and we wish to know if it is reasonable to assume that the population correlation coefficients (ρ) are the same.

TRANSFORMING r TO z_r

Imagine that we have obtained two independent sample r's relating two variables, X and Y, to each other. The sample Pearson r's are

$$r_{xy_1} = 0.83 \quad \text{and} \quad r_{xy_2} = 0.76.$$

We wish to ascertain if it is reasonable to assume that both sample r's were drawn from the same population. The first step is to transform the sample r's into z_r's. Reference to Table D shows that

when $r_1 = 0.83$, $z_{r_1} = 1.188$;

when $r_2 = 0.76$, $z_{r_2} = 0.996$.

Do Exercises 1 through 4 at the right.

CALCULATING THE STANDARD ERROR OF THE DIFFERENCE BETWEEN Z'S (s_{D_z})

The standard error of the difference between z's is

$$s_{D_z} = \sqrt{\frac{1}{n_1 - 3} + \frac{1}{n_2 - 3}}.$$

Examples

If n in sample 1 (n_1) is 20 and n in sample 2 (n_2) is 30, then

$$s_{D_z} = \sqrt{\frac{1}{17} + \frac{1}{27}}$$
$$= \sqrt{0.0588 + 0.0370}$$
$$= 0.31.$$

If $n_1 = 60$ and $n_2 = 75$, then

$$s_{D_z} = \sqrt{\frac{1}{60} + \frac{1}{75}}$$
$$= 0.17.$$

Do Exercises 5 through 8 at the right.

OBJECTIVES

1. Know how to transform sample r's into Fisher's z-statistic.
2. Know how to calculate the standard error of the difference between two z's.
3. Know how to apply the z as the test statistic for evaluation of the null hypothesis, $H_0 : \rho_1 = \rho_2$.

EXERCISES

Transform the following pairs of r's into their respective z's.

1. $r_{xy_1} = -0.15, r_{xy_2} = -0.23$
2. $r_{xy_1} = 0.53, r_{xy_2} = 0.66$
3. $r_{xy_1} = 0.19, r_{xy_2} = 0.11$
4. $r_{xy_1} = -0.35, r_{xy_2} = 0.06$

Calculate s_{D_z} for the following values of n_1 and n_2.

5. $n_1 = 30, n_2 = 33$
6. $n_1 = 57, n_2 = 68$
7. $n_1 = 105, n_2 = 140$
8. $n_1 = 115, n_2 = 110$

5

APPLYING z AS THE TEST STATISTIC

The null hypothesis is that the population correlation coefficients from which the samples were drawn are the same ($H_0: \rho_1 = \rho_2$). The alternative hypothesis is that the population correlation coefficients are different ($H_1: \rho_1 \neq \rho_2$).

The appropriate test of significance is

$$z = \frac{z_{r_1} - z_{r_2}}{s_{Dz}},$$

and the appropriate sampling distribution is the normal probability distribution (Table A).

Example

Given $z_{r_1} = 1.188$, $z_{r_2} = 0.996$, $n_1 = 17$, and $n_2 = 27$, apply z as a test statistic, employing $\alpha = 0.05$, two-tailed test.

$$z = \frac{1.188 - 0.996}{0.31}$$

$$= \frac{0.192}{0.31}$$

$$= 0.62$$

Column C shows that $z = 0.62$ has a two-tailed associated probability of 0.54. Since $0.54 > 0.05$, we cannot reject H_0 at the 5% level of significance.

Do Exercises 9 through 12 at the right.

A Worked Example

Given $r_{xy_1} = 0.92$ and $r_{xy_2} = 0.81$, $n_1 = 60$, $n_2 = 84$, and $\alpha = 0.01$.

Step 1. Convert the sample r's to z_r's. In the present example, $z_{r_1} = 1.589$ and $z_{r_2} = 1.221$.

Step 2. Calculate s_{Dz}, using

$$s_{Dz} = \sqrt{\frac{1}{n_1 - 3} + \frac{1}{n_2 - 3}}.$$

With the present data,

$$s_{Dz} = \sqrt{\frac{1}{57} + \frac{1}{81}}$$

$$= 0.17.$$

Apply the z-statistic and draw the appropriate conclusion in each of the following exercises. Employ $\alpha = 0.05$, two-tailed text.

9. $z_{r_1} = -0.151$, $z_{r_2} = -0.234$
   $n_1 = 30$, $n_2 = 33$

10. $z_{r_1} = 0.590$, $z_{r_2} = 0.739$
    $n_1 = 57$, $n_2 = 68$

11. $z_{r_1} = 0.192$, $z_{r_2} = 0.110$
    $n_1 = 105$, $n_2 = 140$

12. $z_{r_1} = -0.366$, $z_{r_2} = 0.060$
    $n_1 = 15$, $n_2 = 11$

Step 3. Substitute the values found in Steps 1 and 2 in the formula

$$z = \frac{z_{r_1} - z_{r_2}}{s_{D_z}}.$$

In the present example,

$$z = \frac{1.589 - 1.221}{0.17}$$

$$= \frac{0.368}{0.17}$$

$$= 2.16.$$

Step 4. Consult Table A, Column C to find the probability associated with obtained z. If greater than α, do not reject H_0. If equal to or less than α, reject H_0 and assert H_1.

In the present example, the probability associated with $z = 2.16$ is 0.0150. Since $0.0150 > \alpha_{0.01}$, we fail to reject H_0.

Note: The above represents the test of a one-tailed or directional hypothesis. For the test of a nondirectional or two-tailed hypothesis, merely double the p associated with obtained z. In the above example, the two-tailed probability associated with z is $0.0150 \times 2 = 0.0300$.

Do Exercises 13 through 16 at the right.

Given the data in each exercise, test H_0 and draw the appropriate conclusion. Employ $\alpha = 0.05$, two-tailed test.

13. $r_{xy_1} = -0.15, r_{xy_2} = -0.23$
    $n_1 = 30, n_2 = 33$

14. $r_{xy_1} = 0.53, r_{xy_2} = 0.66$
    $n_1 = 57, n_2 = 68$

15. $r_{xy_1} = 0.19, r_{xy_2} = 0.11$
    $n_1 = 105, n_2 = 140$

16. $r_{xy_1} = -0.15, r_{xy_2} = 0.06$
    $n_1 = 15, n_2 = 11$

CHAPTER 5 TEST

To complete this test, students will require access to Table A (Percentage of Areas under the Standard Normal Curve) and Table F (Transformation of r to z_r).

1. Given $r_1 = -0.41$ and $r_2 = -0.21$; their corresponding z_r's are:

   a) 0.436 and 0.214

   b) -0.436 and -0.210

   c) -0.436 and -0.214

   d) 0.390 and -0.214

2. Given $r_1 = 0.53$, $r_2 = -0.62$, and $r_3 = 0.94$; their corresponding z_r's are:

   a) 0.53, -0.725, and 0.74

   b) 0.590, -0.725, and 1.738

   c) 0.49, -0.55, and 1.738

   d) 0.590, 0.725, and -1.738

3. Given $r_1 = 0.58$, $r_2 = 0.73$, $n_1 = 18$, $n_2 = 24$; s_{D_z} equals:

   a) 0.10   b) 0.11   c) 0.31   d) 0.34

4. Given $r_1 = 0.67$, $r_2 = -0.54$, $n_1 = 45$, $n_2 = 52$; s_{D_z} equals:

   a) 0.21   b) 0.20   c) 0.05   d) 0.04

5. Given $r_1 = 0.45$, $r_2 = 0.79$, $s_{D_z} = 0.15$; z equals:

   a) -3.91 b) -3.60 c) 3.60   d) 3.90

6. Given $r_1 = -0.58$, $r_2 = 0.37$, $s_{D_z} = 0.34$; z equals:

   a) 0.74   b) 3.09   c) -0.74 d) -3.09

7. Given $r_1 = 0.86$, $r_2 = 0.49$, $n_1 = 62$, $n_2 = 57$, and $H_0 : \rho_1 \leq \rho_2$. Using $\alpha = 0.01$, we find:

   a) $z = 1.95$, do not reject H_0

   b) $z = 3.98$, do not reject H_0

   c) $z = 3.98$, reject H_0

   d) $z = 1.95$, reject H_0

8. Given $r_1 = 0.71$, $r_2 = -0.11$, $n_1 = 12$, $n_2 = 11$, and $H_0 : \rho_1 \neq \rho_2$. Using $\alpha = 0.05$, we find:

   a) $z = 2.03$, reject H_0

   b) $z = 1.57$, fail to reject H_0

   c) $z = 2.03$, fail to reject H_0

   d) $z = 1.57$, reject H_0

Practical

9. Given $r_{xy_1} = 0.76$, and $r_{xy_2} = 0.38$, $n_1 = 160$, $n_2 = 123$. Using $\alpha = 0.01$, test $H_0 : \rho_1 \leq \rho_2$.

10. Given $r_{xy_1} = 0.27$, and $r_{xy_2} = -0.18$, $n_1 = 54$, $n_2 = 63$. Using $\alpha = 0.05$, test $H_0 : \rho_1 = \rho_2$.

11. Given $r_{xy_1} = 0.24$, and $r_{xy_2} = 0.07$, $n_1 = 416$, $n_2 = 405$. Using $\alpha = 0.01$, test $H_0 : \rho_1 = \rho_2$.

Significance of Difference Between Means, Correlated Samples

There are many factors that contribute to the variability of scores in behavioral research—variations in experimental techniques, differences in capabilities of subjects, and moment-to-moment alterations in such factors as the testing situation, attention and motivation of subjects, etc.   All of these contribute to the variability of the error term.   When these sources of variation are large, the error term is correspondingly large.   It is also less sensitive in evaluating the significance of the difference between means. Research designs employing correlated samples frequently identify and quantify a large source of error and then "remove" this from the error term, to permit a more sensitive evaluation of the difference between and among means.

THE MAIN CLASSES OF CORRELATED SAMPLES DESIGN

There are two main classes of correlated samples design—before-after and matched group designs.

In a before-after design, scores are obtained for each subject both before and after the administration of the experimental conditions. The same or an equivalent scale is used on each occasion.

In a matched group design, subjects in both experimental and control groups are matched in some variable known to be correlated with the criterion task.

Examples

Before-after design :   A test of psychomotor coordination was administered to a group of subjects both before and after the administration of a drug.

Matched group design :   In a study concerned with the evaluation of two different methods of teaching statistics, students were administered a Math Aptitude Scale.   Individuals with similar scores were paired up, with each being assigned at random to a different experimental correlation.

Do Exercises 1 and 2 at the right.

THE STUDENT *t*-RATIO, CORRELATED SAMPLES

The test statistic employed with correlated samples is the Student *t*-ratio,

$$t = \frac{\bar{D} - \mu_D}{s_{\bar{D}}},$$

OBJECTIVES

1.   Know the two main classes of correlated sample designs.

2.   Know how to calculate and interpret the Student *t*-ratio, direct difference method, for correlated samples.

3.   Know how to employ Sandler *A* as an algebraically equivalent substitute for Student *t*, correlated samples design.

EXERCISES

Indicate which of the following employ a before-after design and which a matched group design.

1.   Students were administered a test of mathematics proficiency.   Then they were given a lecture on the value of positive attitudes in test taking.   The test was readministered after the lecture.

2.   On the basis of their weight prior to being administered the experimental conditions, rats were paired in terms of weight and one member of each pair was assigned to each group.

in which

$\bar{D}$ is the difference between the means of the two experimental groups,

μ_D is the difference between means hypothesized under H_0,

$s_{\bar{D}}$ is the standard error of the mean difference.

Examples

If $\bar{D} = 6.4$, $\mu_{\bar{D}} = 2$, and $s_{\bar{D}} = 4.3$, then

$$t = \frac{6.4 - 2}{4.3}$$

$$= \frac{4.4}{4.3}$$

$$= 1.02.$$

Do Exercises 3 and 4 at the right.

The most common null hypothesis is that $\mu_D = 0$. In this event, the Student t-ratio becomes

$$t = \frac{\bar{D}}{s_{\bar{D}}}.$$

Example

Given $H_0 : \mu_D = 0$, $\bar{D} = 8.42$, and $s_{\bar{D}} = 3.72$, then

$$t = \frac{8.42}{3.72}$$

$$= 2.263.$$

Do Exercises 5 and 6 at the right.

The Standard Error of the Mean Difference, $s_{\bar{D}}$

The standard error of the mean difference, $s_{\bar{D}}$, is

$$s_{\bar{D}} = \sqrt{\frac{\sum d^2}{n(n-1)}},$$

in which

$$\sum d^2 = \sum D^2 - \frac{(\sum D)^2}{n},$$

n = the number of pairs, and

$n - 1$ is the number of degrees of freedom.

3. Given

   $\bar{D} = 5.93$, $\mu_{\bar{D}} = 4.00$, and $s_{\bar{D}} = 2.01$;

   find t.

4. Given

   $\bar{D} = -4.44$, $\mu_{\bar{D}} = 2.38$, and $s_{\bar{D}} = 2.04$;

   find t.

5. Given

   $H_0 : \mu_D = 0$, $\bar{D} = -5.73$, and $s_{\bar{D}} = 3.91$;

   find t.

6. Given

   $H_0 : \mu_D = 0$, $\bar{D} = 1.89$, and $s_{\bar{D}} = 0.58$;

   find t.

Example

Given the following sets of paired scores.

Subjects (before-after or matched pairs)	Condition X_1	Condition X_2	D $(X_1 - X_2)$	D^2 $(X_1 - X_2)^2$
A	15	12	3	9
B	19	17	2	4
C	23	25	-2	4
D	30	26	4	16
			$\sum D = 7$	$\sum D^2 = 33$

$$\sum d^2 = 33 - \frac{(7)^2}{4}$$

$$= 33 - 12.25$$

$$= 20.75$$

$$s_{\bar{D}} = \sqrt{\frac{\sum d^2}{n(n-1)}}$$

$$= \sqrt{\frac{20.75}{12}}$$

$$= 1.31.$$

Do Exercises 7 and 8 at the right.

The mean difference, $\bar{D}$, is:

$$\bar{D} = \frac{\sum D}{n}.$$

Example

In the previous example $\sum D = 7$ and $n = 4$. Therefore,

$$\bar{D} = \frac{7}{4} = 1.75.$$

Do Exercises 9 and 10 at the right.

Testing the Null Hypothesis: $H_0 : \mu_D = 0$

The Student t-distributions are the appropriate sampling distributions for testing the null hypothesis when correlated samples are employed.

Find $\sum d^2$ and $s_{\bar{D}}$ in each of the following.

7.

X_1	X_2	D	D^2
15	18		
12	14		
11	13		
9	9		
7	6		

8.

X_1	X_2	D	D^2
25	22		
21	19		
15	18		
14	11		
12	9		

9. Find $\bar{D}$ from the data in Exercise 7.

10. Find $\bar{D}$ from the data in Exercise 8.

6

Example

An investigator speculated that people who smoke tend to smoke more during periods of stress. He compared the number of cigarettes ordinarily smoked by a group of 15 randomly selected students with the number they smoked during the 24 hours prior to final examinations. What did the investigator conclude? Use $\alpha = 0.05$, two-tailed test.

Usual number	Prior to final
8	7
15	10
22	28
11	12
17	19
10	9
6	7
31	39
36	32
18	21
16	18
15	14
17	22
9	9
11	16

1. *Null hypothesis* (H_0): The difference in the mean number of cigarettes usually smoked daily and the number smoked during the 24-hour period preceding a final examination is zero, i.e., $\mu_D = 0$.

2. *Alternative hypothesis* (H_1): The mean number of cigarettes smoked daily is different from the mean number of cigarettes smoked during the 24-hour period preceding a final examination, i.e., $\mu_D \neq 0$.

3. *Statistical test*: Since we are employing a before-after design, the Student t-ratio for correlated samples is appropriate.

4. *Significance level*: $\alpha = 0.05$.

5. *Sampling distribution*: The sampling distribution is the Student t-distribution with df $= n - 1$ (number of pairs minus 1), or $15 - 1 = 14$.

6. *Critical region*: $t \geq 2.145$ and $t \leq -2.145$.

Subject	Usual number	Prior to exam	D	D²
1	8	7	1	1
2	15	10	5	25
3	22	28	−6	36
4	11	12	−1	1
5	17	19	−2	4
6	10	9	1	1
7	6	7	−1	1
8	31	39	−8	64
9	36	32	4	16
10	18	21	−3	9
11	16	18	−2	4
12	15	14	1	1
13	17	22	−5	25
14	9	9	0	0
15	11	16	−5	25
			$\sum D = -21$	$\sum D^2 = 213$

Step 1. The sum of the squares of the difference score $\left(\sum d^2\right)$ is

$$\sum d^2 = \sum D^2 - \frac{\left(\sum D\right)^2}{n}$$

$$= 213 - \frac{(441)}{15}$$

$$= 183.6.$$

Step 2. The standard error of the mean difference is

$$s_{\bar{D}} = \sqrt{\frac{\sum d^2}{n(n-1)}}$$

$$= \sqrt{\frac{183.6}{210}}$$

$$= 0.94.$$

Step 3. The mean difference, $\bar{D}$, is

$$\bar{D} = \frac{\sum D}{n}$$

$$= \frac{-21}{15}$$

$$= -1.40.$$

Step 4. The value of t in the present problem is

$$t = \frac{\bar{D}}{s_{\bar{D}}}$$

$$= \frac{-1.40}{0.94}$$

$$= -1.49.$$

Decision: Since the obtained t does not fall within the critical region (i.e., $-1.49 > t_{0.05}$), we fail to reject H_0.

Do Exercise 11 at the right.

THE SANDLER A-STATISTIC AS AN ALTERNATIVE TO STUDENT'S t-RATIO

The Sandler A-statistic is alegebraically equivalent to the Student t-ratio when correlated samples are employed and H_0 is $\mu_D = 0$.

The A-statistic is defined as follows:

$$A = \frac{\text{the sum of the squares of the differences}}{\text{the square of the sum of the differences}} = \frac{\sum D^2}{(\sum D)^2}.$$

Example

Given $\sum D = 10, \sum D^2 = 50$, then

$$A = \frac{50}{(10)^2} = 0.500.$$

Do Exercises 12 through 14 at the right.

Use of the A-Statistic Table

The number of degrees of freedom is the same as with the Student t-ratio, correlated samples: df $= n - 1$, in which n is the number of pairs.

The critical value, at a given α level, is an obtained A *equal* to *or less* than the tabled value (Table D) for a given number of degrees of freedom.

11. The owner of a small grocery store claimed that on the average, his prices were the same as those at the large neighborhood supermarket. He compared the prices of 10 randomly selected products. What did he conclude? Use $\alpha = 0.05$, two-tailed test.

Item	Grocery	Supermarket	D	D²
1	0.98	0.86		
2	0.23	0.23		
3	0.18	0.18		
4	0.42	0.39		
5	0.57	0.63		
6	0.48	0.49		
7	0.88	0.79		
8	1.33	1.29		
9	1.82	1.75		
10	1.11	0.99		

12. Given $\sum D = 5, \sum D^2 = 120$; find A.

13. Given $\sum D = 40, \sum D^2 = 120$; find A.

14. Given $\sum D = 40, \sum D^2 = 300$; find A.

Example

Given $A = 0.500$, $n = 27$, $\alpha = 0.05$, two-tailed test.

The critical values of the A-statistic are given in Table D. At $\alpha = 0.05$, two-tailed test with df $= 26$, the critical value is 0.265. Since obtained $A = 0.500$ is greater than the tabled critical value of $A = 0.265$, we cannot reject H_0.

Do Exercises 15 through 17 at the right.

A Worked Example

An investigator hypothesizes that the daily caloric intake of single women differs from that of married women. He selects two groups of women, each pair matched for both age and weight, and obtains the following data on caloric intake (data are in hundreds of calories).

Matched pair	Married women	Single women	D	D^2
A	23	20	3	9
B	21	19	2	4
C	20	22	-2	4
D	19	18	1	1
E	19	17	2	4
F	18	15	3	9
G	18	20	-2	4
H	17	17	0	0
I	17	13	4	16
J	17	16	1	1
K	16	14	2	4
L	16	13	3	9
			$\sum D = 17$	$\sum D^2 = 65$

1. *Null hypothesis* (H_0): There is no mean difference in the daily caloric intake of married versus single women, i.e., $\mu_D = 0$.

2. *Alternative hypothesis* (H_1): There is a mean difference in the daily caloric intake of married women versus single women, i.e., $\mu_D \neq 0$.

3. *Statistical test*: Since we are employing a matched group design, the Sandler A-statistic for correlated samples is appropriate.

4. *Significance level*: $\alpha = 0.01$.

5. *Sampling distribution*: The sampling distribution is the Sandler A with df $= n - 1$, or $12 - 1 = 11$.

6. *Critical region*: $A_{0.01} \leq 0.178$.

15. Given $A = 0.480$, $n = 41$, $\alpha = 0.01$, one-tailed test; formulate the statistical decision.

16. Given $A = 0.075$, $n = 31$, $\alpha = 0.01$, two-tailed test; formulate the statistical decision.

17. Given $A = 0.188$, $n = 18$, $\alpha = 0.05$, two-tailed test; formulate the statistical decision.

6

Step 1. Find $\sum D$ and square to obtain $(\sum D)^2$. This value is the denominator in the A-statistic. In the present example, $(\sum D)^2 = (17)^2 = 289$.

Step 2. Square the difference between each matched pair and sum to obtain $\sum D^2$. This value is the numerator in the A-statistic. In the present example, $\sum D^2 = 65$.

Step 3. Divide $(\sum D)^2$ into $\sum D^2$ to obtain A. In the present example

$$A = \frac{65}{289} = 0.225.$$

Since obtained A is less than $A_{0.01}$, we reject H_0 and assert that the daily caloric intake of married women is greater than the caloric intake of their unmarried counterparts.

Do Exercise 18 at the right.

18. The following scores were obtained by twelve pairs of identical twins on a task of motor coordination. Prior to testing, the experimental subjects were administered a psychoactive drug and the controls received a placebo. Set up and test the null hypothesis $\mu_D = 0$, using $\alpha = 0.05$, two-tailed test. Employ Sandler's A-statistic.

Experimental	Control	D	D^2
25	21		
23	20		
21	24		
19	15		
17	20		
17	15		
16	14		
14	10		
14	12		
12	13		
11	6		
10	9		

CHAPTER 6 TEST

To complete this test, students will require access to Table B (Critical Values of t).

1. Given $\sum D = -71$, $n = 12$, $\mu_D = 0.00$, $s_{\bar{D}} = 3.57$; t equals:

   a) -1.66 b) 3.13 c) 1.66 d) -3.13

2. Given $\sum D = 14$, $n = 16$, $\mu_D = 0.00$, and $s_{\bar{D}} = 0.58$; t equals:

   a) 24.14 b) 1.15 c) 1.52 d) 1.72

3. Given $\sum D = 31$, $\sum D^2 = 191$, $n = 11$; $\sum d^2$ equals:

   a) 0.97 b) 10.18 c) 0.94 d) 103.64

4. Given $\sum d^2 = 48.51$, $\sum D = 15$, $n = 8$; $s_{\bar{D}}$ equals:

   a) 4.51 b) 20.38 c) 0.93 d) 0.87

5. Given the following difference scores: $4, 7, 2, -3, 6, 5, 2, -1$; $\bar{D}$ and $\sum d^2$ are:

   a) $3.75; 9.14$ b) $2.75; 9.14$

   c) $2.75; 83.5$ d) $3.75; 83.5$

6. Given $\sum d^2 = 305.62$, $\sum D = 26$, $n = 17$, $H_0 : \mu_D = 0$, $\alpha = 0.01$; we find:

   a) $t = -1.44$, reject H_0 b) $t = 1.37$, reject H_0

   c) $t = 1.37$, fail to reject H_0 d) $t = 1.44$, fail to reject H_0

7. Given $\sum d^2 = 165.40$, $\sum D = 12$, $n = 12$, $H_0 : \mu_D \leq 0.00$, $\alpha = 0.05$; we find:

   a) $t = 0.80$, reject H_0 b) $t = 0.80$, do not reject H_0

   c) $t = 0.89$, reject H_0 d) $t = 0.89$, do not reject H_0

8. Given $\sum D = 24$, $\sum D^2 = 253$, $n = 17$, $\bar{D} = 1.41$; A equals:

   a) 0.8754   b) 0.4392   c) 2.2767   d) 0.0949

9. Given $\sum D = 10$, $\sum D^2 = 203$, $n = 21$, $\bar{D} = 0.48$; A equals:

   a) 2.0300   b) 0.4926   c) 0.0493   d) 0.4603

10. Given the following difference scores: -5, -4, -8, -3, -3, 4, -2, -1, 0. The value of A is:

    a) 0.2975   b) 0.1600   c) 3.3611   d) 6.2500

Practical

11. Given the following scores of matched subjects, test $H_0 : \mu_D \geq 0.00$. Use $\alpha = 0.05$. Find both the Student t-ratio and the Sandler A-statistic.

Experimental condition	Control condition
106	110
103	107
100	105
99	94
96	99
94	101
91	93
90	94
87	90
84	82
82	87
78	84

12. Use the following formula to check the calculation in Question 11:

$$t = \frac{N - 1}{NA - 1}$$

Power of a Statistical Test

As recently as a decade ago, most researchers concentrated on one type of error in the statistical decision-making process, namely, the error of falsely rejecting a true null hypothesis. It is for this reason that the alpha level—which defines the risk of this type of error—is set so low, usually at either the 0.05 or the 0.01 level. The error of falsely rejecting a true null hypothesis is referred to as a type α or type I error. There is another type of error that has attracted more interest in recent years. Referred to as type II or type β (beta) error, it occurs whenever we fail to reject a false null hypothesis. Whereas the probability of a type I error is easily ascertained—it is equal to α—that of a type II error must be calculated. In this chapter, we examine the means of calculating the probability of a type II or type β error and see how the probability of this type of error is related to the statistical power of a test.

THE TWO TYPES OF ERRORS

There are two types of statistical decisions we can make: reject H_0 when the probability of the event-of-interest achieves an acceptable α level (usually $p \leq 0.05$ or ≤ 0.01); fail to reject H_0 when the probability of the event-of-interest is greater than α. With each of these decisions, there is an associated risk of error.

If we reject H_0 (i.e., we conclude H_0 is false) when H_0 is true, we have made the error of falsely rejecting the null hypothesis. This type of error is called a type α or type I error.

If we fail to reject H_0 (i.e., we do not assert the alternative hypothesis), when H_0 is false, we have made the error of falsely accepting H_0. This type of error is referred to as a type β or a type II error.

Examples

a) $H_0: \mu_1 = \mu_2$, $\alpha = 0.05$, two-tailed test. Obtained $p = 0.03$, two-tailed value. Statistical decision: H_0 is false. Actual status of H_0: True.

  Error: Type I—rejecting a true H_0.

b) $H_0: \mu_1 = \mu_2$, $\alpha = 0.05$, two-tailed test. Obtained $p = 0.04$, two-tailed value. Actual status of H_0: False.

  Error: No error has been made. A correct conclusion was drawn since H_0 is false and the statistical decision was that H_0 is false.

c) $H_0: \mu_1 = \mu_2$, $\alpha = 0.01$, two-tailed test. Obtained $p = 0.10$, two-tailed value. Statistical decision: fail to reject H_0. Actual status of H_0: False.

OBJECTIVES

1. Learn to distinguish between type I (type α) and type II (type β) errors.

2. Know the circumstances under which each type of error can occur.

3. Understand the concept of the power of a statistical test.

4. Know how to improve the power of a test.

Error: Type II—failing to reject a false H_0.

d) $H_0: \mu_1 = \mu_2$, $\alpha = 0.01$, two-tailed test. Obtained $p = 0.006$, two-tailed value. Statistical decision: Reject H_0. Actual status of H_0: False.

Error: No error has been made since the statistical decision has been to reject H_0 when H_0 is actually false.

Do Exercises 1 through 5 at the right.

A type I or type α error can only be made when H_0 is true, since this type of error is defined as the mistaken rejection of a true hypothesis. The probability of a type I error is given by the α level.

Example

If $\alpha = 0.01$, the probability of making a type I error is 0.01.

Do Exercise 6 at the right.

A type II or type β error can only be made when H_0 is false, since this type of error is defined as the mistaken acceptance of a false H_0. The probability of making a type II error can be obtained only by calculation.

CALCULATING THE PROBABILITY OF A TYPE II OR TYPE β ERROR

There are two basic steps in calculating the probability of a type II error: (a) finding the critical value for rejecting H_0 under the null distribution and (b) finding the probability of obtaining the critical value—or one more rare—in the "true" sampling distribution of the statistic.

We shall use the two-sample case to illustrate the calculation of the probability of making a type II error. However, we shall assume that the parameters are known and use z as the test statistic. In practice, when parameters are unknown, it is possible to estimate the probability of a type II error by making certain assumptions based on sampling statistics. This latter procedure is beyond the scope of this book.

Given: $\mu_1 = 80$, $\mu_2 = 76$, $\sigma_1^2 = 20$, $\sigma_2^2 = 20$, $n_1 = n_2 = 25$.
$H_0: \mu_1 - \mu_2 = 0$, $H_1: \mu_1 - \mu_2 \neq 0$, $\alpha = 0.05$.

EXERCISES

In each of the following exercises, H_0, α level, obtained p, and true status of H_0 are given. State whether or not an error in statistical decision has been made. If so, state the type of error.

1. $H_0: \mu_1 = \mu_2$, $\alpha \leq 0.01$, one-tailed test. Obtained p is 0.008, one-tailed value (in predicted direction). Actual status of H_0: True.

2. $H_0: \mu_1 = \mu_2$, $\alpha = 0.05$, two-tailed test. Obtained $p = 0.08$, two-tailed value. Actual status of H_0: True.

3. $H_0: \mu_1 = \mu_2$, $\alpha = 0.05$, two-tailed test. Obtained $p = 0.02$, two-tailed value. Actual status of H_0: False.

4. $H_0: \mu_1 = \mu_2$, $\alpha = 0.05$, two-tailed test. Obtained $p = 0.03$, two-tailed value. Actual status of H_0: False.

5. $H_0: \mu_1 = \mu_2$, $\alpha = 0.01$, two-tailed test. Obtained $p = 0.005$, two-tailed value. Actual status of H_0: False.

6. If $\alpha = 0.05$, what is the probability of making a type I error?

Finding the Critical Value for Rejecting H_0 Under the Null Distribution

Step 1.   Set up the problem in formal statistical terms.

a)   *Null hypothesis* (H_0): The samples were drawn from a common population of means so that $\mu_1 - \mu_2 = 0$.

b)   *Alternative hypothesis* (H_1): The sample means were not drawn from a common population of means, that is, $\mu_1 \neq \mu_2$.

c)   *Statistical test*: Since σ_1 and σ_2 are known,

$$z = \frac{\bar{X}_1 - \bar{X}_2}{\sigma_{\bar{X}_1 - \bar{X}_2}}$$

is the appropriate test statistic.

d)   *Significance level*: $\alpha = 0.05$, two-tailed test.

e)   *Sampling distribution*: The normal curve.

f)   *Critical region*: $z \leq -1.96$ or ≥ 1.96.

Therefore the critical value of the difference in sample means (the minimum value of $\bar{X}_1 - \bar{X}_2$ leading to rejection of H_0) is

critical value: $\bar{X}_1 - \bar{X}_2 = \pm 1.96\sigma_{\bar{X}_1 - \bar{X}_2}$.

Step 2.   Calculate the value of $\sigma_{\bar{X}_1 - \bar{X}_2}$:

$$\sigma_{\bar{X}_1 - \bar{X}_2} = \sqrt{\frac{\sigma_1^2}{n_1} + \frac{\sigma_2^2}{n_2}} = \sqrt{\frac{20}{25} + \frac{20}{25}} = 1.26.$$

Step 3.   Find the critical value of the difference between means:

critical value: $\bar{X}_1 - \bar{X}_2 = \pm 1.96(1.26)$

$$= \pm 2.47.$$

Thus any difference equal to or greater than 2.47 or equal to or less than -2.47 will lead to the rejection of H_0 under the null distribution.

Do Exercises 7 through 9 at the right.

Finding the Probability of Obtaining the Critical Value in the True Sampling Distribution of the Statistic

Recall that the difference between means in the true sampling distribution of the statistic is $80 - 76 = 4.00$.

Given

$\mu_1 = 40$, $\mu_2 = 35$, $\sigma_1^2 = 15$, $\sigma_2^2 = 15$, $n_1 = n_2 = 20$; $H_0: \mu_1 \leq \mu_2$, $H_1: \mu_1 > \mu_2$, $\alpha = 0.01$.

7.   Set this problem up in formal statistical terms.

8.   Calculate the value of $\sigma_{\bar{X}_1 - \bar{X}_2}$.

9.   Find the critical value for rejecting H_0 under the null distribution.

Step 4. The z corresponding to the negative minimal value of the difference between means in the true sampling distribution is

$$z = \frac{-2.47 - 4.00}{\sigma_{\bar{X}_1 - \bar{X}_2}}$$

$$= \frac{-2.47 - 4.00}{1.26}$$

$$= -5.13.$$

The probability associated with this value of z is so small that we shall assume it is equal to zero. In other words, the probability is essentially zero that we will obtain a test statistic this deviant in the true distribution of differences between means. We may safely ignore this possibility, as seen in the accompanying diagram.

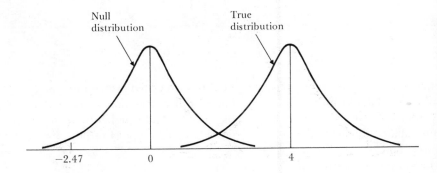

Step 5. The z corresponding to the positive minimal value of the difference between means in the true sampling distribution is

$$z = \frac{2.47 - 4}{\sigma_{\bar{X}_1 - \bar{X}_2}}$$

$$= \frac{-1.53}{1.26}$$

$$= -1.21.$$

Any obtained z equal to or less than -1.21 will lead to the false acceptance of H_0 (a type II or type β error). Reference to Table A shows the area beyond $z = -1.21$ is 0.1131. In other words, the probability is only 0.1131 of making a type II error when drawing two samples of $n = 25$ from populations in which $\mu_1 - \mu_2 = 4$ and $\sigma_1^2 = \sigma_2^2 = 20$.

The accompanying diagram shows the area corresponding to difference in sample statistics which would not lead to rejection of H_0.

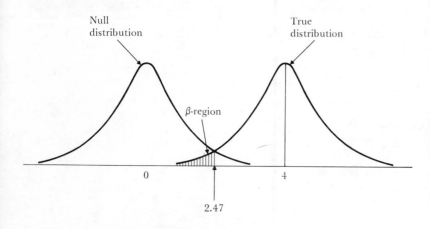

Do Exercises 10 and 11 at the right.

THE CONCEPT OF POWER

When we calculated the probability of making a type II error, we determined the probability that we would fail to reject a false null hypothesis. By subtracting this β probability from 1.00, we ascertain the power of a test. Power is defined as the probability of making the correct decision when H_0 is false—that is, power is the probability of rejecting a false null hypothesis. In the previous example, in which β probability was found to be 0.1131, the power of the test is

power $= 1.00 - \beta$

$= 1.00 - 0.1131$

$= 0.8869.$

In the parlance of statistics, this is a very powerful test since it would lead to the correct decision almost 89% of the time.

Do Exercise 12 at the right.

IMPROVING THE POWER OF A TEST

There are several factors that influence the power of a test including sample size, α level, the directionality of H_0 and H_1, and the precision of estimating experimental error.

For Exercises 10 and 11, use the same facts as in Exercises 7 through 9.

10. Find the z's corresponding to both the negative and the positive minimal values of the differences between means in the true sampling distribution.

11. Find the associated probabilities of finding differences more extreme than the above. Add together to obtain the probability of a type II error.

12. Calculate the power of the test that appears in Exercises 7 through 11.

7

Sample Size and Power

In the previous example, when $n_1 = n_2 = 25$, the power of the test was found to be 0.8869.

If we had used a smaller n, as for example $n_1 = n_2 = 10$, the power of the test would have been

$$\text{power} = 1 - \beta$$
$$= 1 - 0.4840$$
$$= 0.5160.$$

If we had used a larger n, as for example $n_1 = n_2 = 40$, the power of the test would have been

$$\text{power} = 1 - \beta$$
$$= 1 - 0.0207$$
$$= 0.9793.$$

Do Exercises 13 through 16 at the right.

Alpha (α) Level and Power

As the α level is decreased, we decrease the probability of a type I error and increase the probability of a type II error. Conversely, as the α level is increased, we increase the probability of a type I error and decrease the probability of a type II error.

Since the power of a test is inversely related to the probability of a type II error (i.e., power increases as the probability of a type II error decreases), it follows that the power can be increased by setting a higher α level for rejecting H_0.

Example

In the previous example, in which $\mu_1 = 80$, $\mu_2 = 76$, $\sigma_1^2 = 20$, $\sigma_2^2 = 20$, $n_1 = n_2 = 25$, $H_0: \mu_1 - \mu_2 = 0$, $H_1: \mu_1 - \mu_2 \neq 0$, and $\alpha = 0.05$, we found the power to be 0.8869.

Had we used $\alpha = 0.01$, the critical value of the difference between means would have been

$$\text{critical value}: \bar{X}_1 - \bar{X}_2 = (2.58)(1.26)$$
$$= 3.25.$$

The probability of a type II error would have been 0.2743 and the power would have been equal to 0.7257.

---

13. Confirm that the power of the test would have been 0.5160 if an n of 10 had been used in each group.

14. Confirm that the power of the test would have been 0.9793 if an n of 40 had been used in each group.

15. For the present problem, plot a graph of the power of the test as a function of N ($N = 20$, $N = 50$, $N = 80$).

16. Generalize: What is the effect of sample size on power?

Do Exercise 17 at the right.

The Directionality of H_0 and H_1

The effect of using directional hypotheses is the same as lowering the α level.

Example

In the previous problem, the critical value of z at $\alpha = 0.05$, one-tailed test, would have been 1.65. Consequently, the critical value would have been

$$\text{critical value: } \bar{X}_1 - \bar{X}_2 = (1.65)(1.26)$$
$$= 2.08.$$

The probability of a type II error would have been 0.0188 and the power of the test would have been equal to 0.9812.

The test of a one-tailed hypothesis is more powerful than its two-tailed counterpart *only when the parameter or parameters are in the predicted direction.*

Do Exercise 18 at the right.

The Precision of Estimating Experimental Error

Any factor that increases the precision in our measurement of experimental error will also increase the power of a test. The reason is that the increased precision will reduce $\sigma_{\bar{X}_1 - \bar{X}_2}$, which will in turn decrease the magnitude of the critical value required for rejecting H_0. There are numerous ways of increasing the precision of our measurement of experimental error, including: improving the reliability of our criterion measure, standardizing the experimental technique, and using correlated measures. In a matched pairs design, for example, the power of the test will increase as the correlation between paired measures increases.

Do Exercise 19 at the right.

17. Confirm that the power of the test would have been 0.7257 if $\alpha = 0.01$ had been used.

18. Confirm that the power of the test would have been 0.9812 if a one-tailed test at $\alpha = 0.05$ had been used.

19. All other factors being equal except the correlation between the criterion scores of matched groups, which of the following would yield the highest power? (a) $r = 0.14$, (b) $r = -0.53$, (c) $r = 0.81$, (d) $r = 0.00$.

CHAPTER 7 TEST

To complete this test, students will require access to Table A (Percentage of Areas under the Standard Normal Curve).

1. Given $H_0 : \mu_1 \geq \mu_2$, $\alpha = 0.05$, one-tailed test. Obtained p is 0.04, in predicted direction, one-tailed value. Actual status of H_0: False.

   a) We accept H_0 and make a type II error.

   b) We reject H_0 and make a type α error.

   c) We accept H_0 and make the correct decision.

   d) We reject H_0 and make the correct decision.

2. Given $H_0 : \mu_1 = \mu_2$, $\alpha = 0.01$, two-tailed test. Obtained p is 0.008, two-tailed value. Actual status of H_0: True.

   a) We accept H_0 and make a type II error.

   b) We reject H_0 and make a type α error.

   c) We accept H_0 and make the correct decision.

   d) We reject H_0 and make the correct decision.

3. Given $H_0 : \mu_1 = \mu_2$, $\alpha = 0.05$, two-tailed test. Obtained p is 0.40, two-tailed value. Actual status of H_0: False.

   a) We accept H_0 and make a type II error.

   b) We reject H_0 and make a type α error.

   c) We accept H_0 and make the correct decision.

   d) We reject H_0 and make the correct decision.

4. Given $H_0 : \mu_1 = \mu_2$, $\alpha = 0.01$, two-tailed test. Obtained p is 0.10, two-tailed value. Actual status of H_0: True.

   a) We accept H_0 and make a type II error.

   b) We reject H_0 and make a type α error.

   c) We accept H_0 and make the correct decision.

   d) We reject H_0 and make the correct decision.

5. If β equals 0.2650, two-tailed test, the probability of making a type II error is:

   a) 0.2650   b) 0.7350

   c) $1 - \beta$ d) insufficient information

6. Given $\sigma_1^2 = 30$, $n_1 = n_2 = 16$, $H_0 : \mu_1 = \mu_2$, $\alpha = 0.05$, two-tailed test; $\sigma_{\bar{x}_1 - \bar{x}_2}$ equals:

   a) 3.75   b) 1.37   c) 1.94   d) 1.03

7. Given $n_1 = n_2 = 60$, $H_0 : \mu_1 \leq \mu_2$, $\alpha = 0.05$, one-tailed test, and $\sigma_{\bar{X}_1 - \bar{X}_2} = 2.40$; critical value $\bar{X}_1 - \bar{X}_2$ equals:

   a) 4.70   b) 5.59   c) 6.19   d) 3.96

8. If the z's corresponding to the negative and positive minimal values of the difference between means in the true sampling distribution are -5.05 and -1.03, the probability of a type II error is:

   a) 0.8508   b) 0.6492   c) 0.3508   d) 0.1492

9. If $\alpha = 0.05$ and the probability of a type II error is 0.1457, the power of the test is:

   a) 0.95   b) 0.8543   c) 0.6457

   d) The concept of power does not apply.

10. Which of the following will *not* increase the power of a test?

   a) reducing experimental error

   b) decreasing α-level

   c) increasing N

   d) using directional hypotheses where appropriate

Practical

11. Given two normal populations in which $\mu_1 = 15$, $\mu_2 = 12$, $\sigma_1 = 3$, $\sigma_2 = 3$.

   a) Determine the power of the test when $\alpha = 0.05$, two-tailed value, and $n_1 = n_2 = 40$.

   b) Determine the power of the test when $\alpha = 0.01$, two-tailed value, and $n_1 = n_2 = 20$.

   c) Determine the power of the test when $\alpha = 0.05$, two-tailed value, and $n_1 = n_2 = 20$.

   d) Determine the power of the test when $\alpha = 0.01$, two-tailed value, and $n_1 = n_2 = 40$.

One-Way Analysis of Variance, Independent Samples

In a one-way analysis of variance, there is only one experimental or treatment variable. This variable may be either qualitative (e.g., method of instructing math to third graders, type of psychotherapy, method of dispersing welfare payments, or type of penal institution for handling prisoners convicted of violent crimes) or quantitative (e.g., amount of drug administered to a group of psychiatric patients, amount of fertilizer applied to a given crop, or amount of positive reinforcement administered to subjects in a learning situation).

While there is only one experimental variable in a one-way analysis of variance, there may be any number of subclasses or levels of treatment. For example, qualitative variables might involve four methods of teaching math, eight types of psychotherapy, four methods of dispensing welfare payments, or five types of penal institutions for handling prisoners convicted of violent crimes. Quantitative variables may involve four dosage levels or amounts of drug administered to patients, five different amounts of fertilizer applied to crops, or three different amounts of positive reinforcement administered to subjects in a learning situation.

Do Exercises 1 through 5 at the right.

BETWEEN-GROUP AND WITHIN-GROUP VARIANCE ESTIMATES

The one-way analysis of variance permits the researcher to collect data on several treatment groups simultaneously and evaluate the null hypothesis that the sample means were drawn from a common population of means. For example, in a five-group design, the null hypothesis is

$$H_0: \mu_1 = \mu_2 = \mu_3 = \mu_4 = \mu_5.$$

The alternative hypothesis is that the samples were not all drawn from the same population of means. If H_0 is rejected, H_1 may be asserted—the researcher may feel confident that there is a treatment effect.

The test statistic (the F-ratio) is the ratio between two variance estimates, one based on the dispersion of scores within groups (within-group variance estimates) and the other based on the dispersion of means (between-group variance estimate). In the one-way analysis of variance, independent group design,

$$F = \frac{\text{between-group variance estimate}}{\text{within-group variance estimate}} = \frac{s_B^2}{s_W^2}.$$

OBJECTIVES

1. Know the basic characteristics of a one-way analysis of variance, independent samples design.

2. Distinguish between one-way analyses of variance employing qualitative and quantitative variables.

3. Know the relationship among the total sum of squares, between-group sum of squares, and within-group sum of squares.

4. Know how to obtain variance estimates and to apply the F-ratio to evaluate the null hypothesis.

EXERCISES

Indicate which is a qualitative and which is a quantitative treatment variable.

1. Amount of negative reinforcement.

2. Level of fear arousal.

3. Type of fertilizer.

4. Amount of punishment.

5. Type of punishment.

If the between-group variance estimate is large (that is, the differences among means are large) relative to the within-group variance estimate (the differences among scores are relatively small), the F-ratio is large. In general, the larger the F-ratio, the greater the likelihood that H_0 will be rejected.

Do Exercises 6 through 10 at the right.

Another way of stating the null hypothesis is that the between-group variance estimates and the within-group variance estimates are both estimates of random error (σ_ε^2).

When H_0 is true and there is no treatment effect, the between-group variance estimate will provide an estimate of σ_ε^2, or random error.

The within-group variance, which contains no treatment effect, is also estimating random error (σ_ε^2). Thus, when H_0 is true, the F-ratio would consist of

$$F = \frac{\sigma_\varepsilon^2}{\sigma_\varepsilon^2}.$$

Allowing for random variations in error estimates from sample data, the expected value of F is 1.00 when H_0 is true.

When H_0 is false, however, the between-group sum of squares contains both treatment effects (σ_β^2) and random error (σ_ε^2). Thus, when H_0 is false, the F-ratio consists of

$$F = \frac{\sigma_\varepsilon^2 + \sigma_\beta^2}{\sigma_\varepsilon^2}.$$

The greater the treatment effect or the larger the between-group sum of squares, the greater the F-ratio becomes.

Do Exercises 11 through 15 at the right.

PARTITIONING THE SUM OF SQUARES

The total sum of squares consists of the sum of the squared deviations of each score from the mean of all scores:

$$\sum x_T^2 = \sum (X - \bar{X}_T)^2.$$

6. A researcher conducts a study in which there are eight different treatment groups. Set up the null hypothesis. Formulate the alternative hypothesis.

7. A researcher conducts a study in which there are four different treatment groups. Set up the null hypothesis and formulate the alternative hypothesis.

8. If the differences among means are small relative to the differences among scores, the F-ratio will be large/small? (Select one.)

9. If the between-group variance estimate is large relative to the within-group variance estimate, the F-ratio will be large/small? (Select one.)

10. In general, the between-group variance estimate will be large if the differences among means are large/small? (Select one.)

Find the F-ratio for each of the following.

11. $\sigma_\varepsilon^2 = 20,\ \sigma_\beta^2 = 35$

12. $\sigma_\varepsilon^2 = 10,\ \sigma_\beta^2 = 0$

13. $\sigma_\varepsilon^2 = 40,\ \sigma_\beta^2 = 35$

14. $\sigma_\varepsilon^2 = 3,\ \sigma_\beta^2 = 0$

15. $\sigma_\varepsilon^2 = 20,\ \sigma_\beta^2 = 80$

Example

Condition A_1	$(X - \bar{X}_T)$	$(X - \bar{X}_T)^2$
10	4	16
8	2	4
6	0	0
4	-2	4
$\sum A_1 = 28$		$\sum(X - \bar{X}_T)^2 = 24$

Condition A_2	$(X - \bar{X}_T)$	$(X - \bar{X}_T)^2$
8	2	4
6	0	0
4	-2	4
2	-4	16
$\sum A_2 = 20$		$\sum(X - \bar{X}_T)^2 = 24$

$$\bar{X}_T = \frac{\sum A_1 + \sum A_2}{n_1 + n_2} = \frac{48}{8} = 6.00$$

$$\sum x_T^2 = 24 + 24 = 48$$

Do Exercise 16 at the right.

The within-group sum of squares consists of the pooled sum of squares within each condition, i.e.,

$$\sum x_W^2 = \sum x_1^2 + \sum x_2^2 + \cdots + \sum x_k^2,$$

in which k is the kth group.

Example

For the above data:

Condition A_1	$(X - \bar{X}_{A_1})$	$(X - \bar{X}_{A_1})^2$
10	3	9
8	1	1
6	-1	1
4	-3	9
$\sum A_1 = 28$		$\sum(X - X_{A_1})^2 = 20$
	$\bar{X}_{A_1} = 7$	

16. Find the total sum of squares $\left(\sum x_T^2\right)$ for the following set of data:

A_1	A_2
15	12
12	9
9	6
6	3
3	0

Condition A_2	$(X - \bar{X}_{A_2})$	$(X - \bar{X}_{A_2})^2$
8	3	9
6	1	1
4	−1	1
2	−3	9
$\sum A_2 = 20$		$\sum (X - \bar{X}_{A_2})^2 = 20$
	$\bar{X}_{A_2} = 5$	

$$\sum x_W^2 = \sum x_{A_1}^2 + \sum x_{A_2}^2$$
$$= 20 + 20$$
$$= 40$$

Do Exercise 17 at the right.

The between-group sum of squares consists of the sum of the squared differences between each group mean $(\bar{X}_B)$ and the total mean $(\bar{X}_T)$, weighted for the n in each condition:

$$\sum x_B^2 = \sum n_i (\bar{X}_i - \bar{X}_T)^2,$$

in which

n_i is the number in the ith group, and
$\bar{X}_i$ is the mean of the ith group.

Example

For the preceding data,

$\bar{X}_{A_1} = 7, \bar{X}_{A_2} = 5, \bar{X}_T = 6$

$n_1 = 4, n_2 = 4$

$$\sum x_B^2 = 4(7 - 6)^2 + 4(5 - 6)^2$$
$$= 4 + 4$$
$$= 8.$$

Do Exercise 18 at the right.

Note that the total sum of squares $(\sum x_T^2)$ has been partitioned into two components, the between-group sum of squares $(\sum x_B^2)$ and the within-group sum of squares $(\sum x_W^2)$.

The sum of these components equals the total sum of squares:

$$\sum x_T^2 = \sum x_B^2 + \sum x_W^2.$$

17. Find the within-group sum of squares $(\sum x_W^2)$ for the data in Exercise 16.

18. Find the between-group sum of squares $(\sum x_B^2)$ for the data in Exercise 16.

In the sample problem we saw that

$$\sum x_T^2 = 48, \qquad \sum x_B^2 = 8, \qquad \sum x_W^2 = 40.$$

Thus

$$\sum x_T^2 = \sum x_B^2 + \sum x_W^2,$$

$$48 = 8 + 40.$$

Do Exercise 19 at the right.

OBTAINING VARIANCE ESTIMATES

We are interested in obtaining two variance estimates: s_W^2 (within-group or error variance) and s_B^2 (between-group variance). The variance estimates are obtained by dividing the sum of squares by the appropriate number of degrees of freedom, i.e.,

$$s_W^2 = \frac{\sum x_W^2}{\mathrm{df}} \qquad \text{and} \qquad s_B^2 = \frac{\sum x_B^2}{\mathrm{df}}.$$

The number of within-group degrees of freedom is $N - k$ (the total number of observations minus the number of treatment conditions). In the previous example $N = 8$ and $k = 2$. Therefore, $\mathrm{df}_W = 8 - 2 = 6$.

The number of between-group degrees of freedom is $k - 1$ (the number of treatment conditions minus 1). In the preceding example, $k = 2$. Therefore, $\mathrm{df}_B = 2 - 1 = 1$.

The total number of degrees of freedom (df_T) is equal to $N - 1$. In the preceding example, $N = 8$. Therefore $\mathrm{df}_T = 8 - 1 = 7$. Note that

$$\mathrm{df}_T = \mathrm{df}_W + \mathrm{df}_B,$$

$$7 = 6 + 1.$$

Do Exercise 20 at the right.

We may now obtain the two variance estimates, s_W^2 and s_B^2, for the preceding example:

$$s_W^2 = \frac{\sum x_W^2}{\mathrm{df}_W} = \frac{40}{6} = 6.67,$$

$$s_B^2 = \frac{\sum x_B^2}{\mathrm{df}_B} = \frac{8}{1} = 8.00.$$

19. Show for the data in Exercises 16 through 18 that
$$\sum x_T^2 = \sum x_B^2 + \sum x_W^2.$$

20. For the data in Exercise 16, find (a) df_T, (b) df_W, (c) df_B.

8

Do Exercise 21 at the right.

RAW SCORE FORMULAS FOR TOTAL, BETWEEN-GROUP, AND WITHIN-GROUP SUM OF SQUARES

A Worked Example

A manufacturer selects a number of samples of 100 items per sample from three different processes used in manufacturing the items. The following number of defective units was found in the various samples that were taken. Determine if H_0 is tenable that the sample means are drawn from a common population of means. Use $\alpha = 0.05$.

| Process 1 | | Process 2 | | Process 3 | |
X_1	X_1^2	X_2	X_2^2	X_3	X_3^2
2	4	7	49	8	64
2	4	3	9	4	16
7	49	7	49	5	25
2	4	9	81	9	81
5	25	4	16	10	100
4	16	5	25	11	121
3	9	3	9	9	81
$\sum X_1 = 25$ $\sum X_1^2 = 111$		$\sum X_2 = 38$ $\sum X_2^2 = 238$		$\sum X_3 = 56$ $\sum X_3^2 = 488$	

Step 1. Find $\sum X$ for each condition. Add these together to find $\sum X_T$. In the present problem,

$$\sum X_T = \sum X_1 + \sum X_2 + \sum X_3$$
$$= 25 + 38 + 56$$
$$= 119.$$

Step 2. Square $\sum X_T$ and divide by N. With the present data,

$$\frac{(\sum X_T)^2}{N} = \frac{(119)^2}{21} = 674.33.$$

Step 3. Square each score in each condition and sum to obtain $\sum X^2$ for each group. Add these together to obtain $\sum X_T^2$. In the present problem,

$$\sum X_T^2 = \sum X_1^2 + \sum X_2^2 + \sum X_3^2$$
$$= 111 + 238 + 488$$
$$= 837.$$

21. Obtain the following variance estimates for the data in Exercise 16: (a) s_W^2, (b) s_B^2.

Step 4. Find the total sum of squares by subtracting the value found in Step 2 from the sum found in Step 3:

$$\sum x_T^2 = \sum X_T^2 - \frac{(\sum X_T)^2}{N}$$

$$= 837 - 674.33$$

$$= 162.67.$$

Step 5. Find the total number of degrees of freedom ($\text{df}_T = N - 1$):

$$\text{df}_T = N - 1$$

$$= 21 - 1$$

$$= 20.$$

Step 6. Place the values found in Steps 4 and 5 in the summary table found in Step 13.

Step 7. Find the between-group sum of squares by squaring each $\sum X$, dividing by the n in the condition, and summing. Subtract the value found in Step 2 from this sum. In the present problem,

$$\sum x_B^2 = \frac{(\sum X_1)^2}{n_1} + \frac{(\sum X_2)^2}{n_2} + \frac{(\sum X_3)^2}{n_3} - 674.33$$

$$= \frac{(25)^2}{7} + \frac{(38)^2}{7} + \frac{(56)^2}{7} - 674.33.$$

Since the n's are equal, there will be less rounding error if we put all the quantities to be squared over the least common denominator (7):

$$\sum x_B^2 = \frac{(25)^2 + (38)^3 + (56)^2}{7} - 674.33$$

$$= \frac{5205}{7} - 674.33$$

$$= 69.24.$$

Step 8. Find $\text{df}_B = k - 1$. In the present problem, the number of groups (k) is equal to 3. Thus df $= 3 - 1 = 2$.

Step 9. Place the values found in Steps 7 and 8 in the summary table found in Step 13.

Step 10. Find the within-group sum of squares. This may be obtained by subtracting, since

$$\sum x_W^2 = \sum x_T^2 - \sum x_B^2$$

$$= 162.67 - 69.24$$

$$= 93.43.$$

8

However, if you have made an error in calculating $\sum x_T^2$ or $\sum x_B^2$, the error may go undetected. It is recommended that you obtain $\sum x_W^2$ as a check of the accuracy of your calculations. The following formula is employed to obtain $\sum x_W^2$ directly:

$$\sum x_W^2 = \sum X_1^2 - \frac{(\sum X_1)^2}{n_1} + \sum X_2^2 - \frac{(\sum X_2)^2}{n_2} + \sum X_3^2 - \frac{(\sum X_3)^2}{n_3}$$

$$= 111 - \frac{(25)^2}{7} + 238 - \frac{(38)^2}{7} + 488 - \frac{(56)^2}{7}$$

$$= 21.71 + 31.71 + 40.00$$

$$= 93.42.$$

(The disparity of 0.01 between the two calculations of $\sum x_W^2$ represents rounding error.)

Step 11. Find $df_W = N - k$. In the present problem, the number of groups (k) is equal to 3. Thus

$$df_W = 21 - 3 = 18.$$

Step 12. Place the value found in Steps 10 and 11 in the summary table found in Step 13.

Step 13. Summary table for representing the relevant statistics in analysis-of-variance problems:

Source of variation	Sum of squares	Degrees of freedom	Variance estimate	F
Between groups	69.24	2	34.62	
Within groups	93.43	18	5.19	
Total	162.67	20		

Step 14. Check to ascertain that

$$\sum x_T^2 = \sum x_B^2 + \sum x_W^2$$

and

$$df_T = df_B + df_W.$$

Step 15. Find the between-group variance estimate by dividing the between-group sum of squares by df_B. In the present problem,

$$s_B^2 = \frac{69.24}{2} = 34.62.$$

Place in the summary table shown in Step 13.

Step 16. Find the within-group variance estimate by dividing the within-group sum of squares by df_W:

$$s_W^2 = \frac{93.43}{18} = 5.19$$

Place in the table shown in Step 13.

Step 17.   Find the *F*-ratio by the following:

$$F = \frac{s_B^2}{s_W^2} = \frac{34.62}{5.19}$$

$$= 6.67, \qquad \text{df } 2/18.$$

Step 18.   Look up $F = 6.67$ in Table C, under 2 and 18 df. If $F = 6.67$ equals or exceeds the tabled value at $\alpha = 0.05$ (lightface type) we reject H_0 and assert that the processes do not represent a common population of means.

Since $F = 6.67 > 4.38$, we reject H_0. Apparently the processes produce different numbers of defective items.

Do Exercises 22 and 23 at the right.

CHAPTER 8 TEST

1. Which of the following is a quantitative variable?

   a)   time to react to a stimulus   b)   type of vegetable

   c)   type of furniture   d)   city of residence

2. In a one-way analysis of variance:

   a)   total sum of squares may be partitioned into only one source of variation

   b)   there may be any number of levels of a sample variable

   c)   between-group sum of squares minus within-group sum of squares equals total sum of squares

   d)   none of the preceding.

3. In a one-way analysis of variance:

   a) $\sum x_{T_2}^2 + \sum x_{W_2}^2 = \sum x_B^2$ b) $\sum x_W^2 + \sum x_B^2 = \sum x_T^2$

   c) $\sum x_B^2 - \sum x_T^2 = \sum x_W^2$ d) $\sum x_W^2 - \sum x_T^2 = \sum x_B^2.$

4. When the *F*-ratio is large:

   a)   the differences among means are large when compared to the differences among scores

   b)   the differences among means are small when compared to the differences among scores

22. Conduct an analysis of variance of the following data, employing the raw score formulas.   Use $\alpha = 0.05$.

X_1	X_2
10	8
8	6
6	4
4	2

23. Conduct an analysis of variance of the following data, using the raw score formulas.   Use $\alpha = 0.01$.

X_1	X_2	X_3	X_4	X_5
13	20	17	14	12
19	10	15	11	5
16	15	12	9	7
21	17	11	5	10
34	19	8	8	6
17	14	11	10	4

c)   the differences among both means and scores are large

d)   the differences among both means and scores are small

5. When H_0 is true, the expected value of F is:

a)   less than zero

b)   greater than zero but less than 1.00

c)   equal to 1.00

d)   greater than 1.00

6. Given $\sigma_\varepsilon^2 = 105$, $\sigma_B^2 = 145$; F equals:

   a)   1.38   b)   0.72   c)   0.58   d)   2.38

7. Given $\sum x_W^2 = 154$, $\sum x_T^2 = 204$; $\sum x_B^2$ equals:

   a)   −50   b)   358   c)   50   d)   1.32

8. Given $\sum X_T^2 = 1565$, $\sum X_T = 215$, $N_T = 36$; $\sum x_T^2$ equals:

   a)   280.97   b)   43.47   c)   269.00   d)   1350.00

9. Given $\sum X_1 = 27$, $\sum X_2 = 20$, $\sum X_3 = 10$, $n_1 = n_2 = n_3 = 8$;
   $\sum x_B^2$ equals:

   a)   1229.00   b)   153.63   c)   18.26   d)   406.13

10. Given $N_T = 90$, $n_1 = n_2 = n_3 = n_4 = n_5 = n_6 = 15$; df_W equals:

    a)   5   b)   84   c)   14   d)   89

11. Given $N_T = 48$, $n_1 = n_2 = n_3 = n_4 = 12$; df_B equals:

    a)   11   b)   47   c)   44   d)   3

12. Given $N_T = 51$, $n_1 = n_2 = n_3 = 17$; df_T equals:

    a)   50   b)   2   c)   48   d)   16

13. Given $\sum x_T^2 = 219$, $\sum x_B^2 = 105$, $\sum x_W^2 = 114$, $df_T = 74$, $df_B = 4$,
    $df_N = 70$; F equals:

    a)   0.92   b)   0.48   c)   14.72   d)   8.87

14. Given $s_B^2 = 65$, $s_W^2 = 14$, $df_B = 4$, $df_W = 30$, $\alpha = 0.01$; we conclude:

    a) $F = 4.64$, reject H_0 b) $F = 34.57$, fail to reject H_0

    c) $F = 4.64$, fail to reject H_0 d) $F = 34.57$, reject H_0

15. Given $\sum x_T^2 = 196$, $\sum x_W^2 = 146$, $N_T = 27$, $k = 3$, $\alpha = 0.05$; we
    conclude:

    a) $F = 4.11$, fail to reject H_0 b) $F = 4.11$, reject H_0

    c) $F = 0.34$, reject H_0 d) $F = 0.34$, fail to reject H_0.

Practical

16. Conduct an analysis of variance of the following set of scores made
    by subjects in four treatment groups of a single experimental
    variable. Use $\alpha = 0.01$ in evaluating the outcome of the experiment.

X_1	X_2	X_3	X_4
6	1	10	9
8	6	6	7
1	9	9	6
4	6	8	3
3	7	4	2
9	5	4	9
5	7	1	8
1	5	7	6

Multicomparison Tests

In Chapter 8 we demonstrated a one-way analysis of variance in which there were three treatment groups. The fact that we obtained a statistically significant F-ratio permits us to conclude that the treatment means were not drawn from a common population of means. However, our research interest usually extends beyond the demonstration of an *overall* treatment effect. We often want to know where the *specific* differences lie, i.e., which of the means differ significantly from one another? To answer this question, a number of different tests have been developed. In this chapter we illustrate the use of one of the more widely accepted multicomparison tests—the Tukey HSD (Honestly Significant Difference) test.

PLANNED (*A PRIORI*) VS. UNPLANNED (*A POSTERIORI*) COMPARISONS

Occasionally a researcher will plan, *in advance of the conduct of the research*, specific comparisons in which he or she has an interest. Such comparisons are referred to as *planned* or *a priori* comparisons. The F-ratio need not be significant in order to justify making these planned comparisons.

In contrast, many studies are designed in which the comparisons of interest are not specified in advance. Although the researcher expects there to be treatment effects, he or she is unable or unwilling to specify what these effects will be. If the F-ratio is statistically significant, the researcher is then free to investigate specific comparisons. Multicomparison tests administered after the fact are referred to as unplanned comparisons, *a posteriori* comparisons, or *post hoc* tests.

In this text, we shall illustrate one multicomparison test—the Tukey HSD test. The HSD test involves *a posteriori* comparisons.

Do Exercises 1 and 2 at the right.

THE TUKEY HSD (HONESTLY SIGNIFICANT DIFFERENCE) TEST

Number of Pairwise Comparisons

The Tukey HSD test may be used to make pairwise comparisons between treatment means. The number of pairwise comparisons for k treatment conditions is given by

$$\frac{k(k-1)}{2}.$$

EXERCISES

Indicate the type of comparison—planned or unplanned—in each exercise.

1. Investigator A conducts a one-way analysis of variance with four treatment conditions. Finding a significant F-ratio, he decides to compare the mean of each condition with the mean of every other condition.

2. Investigator B also conducts a study involving four treatment levels of a single variable. Prior to the study, however, she hypothesizes that the combined treatments A_1 and A_2 will have higher means than the combined treatments A_3 and A_4.

9

Examples

If there are four treatment conditions, the number of pairwise comparisons is

$$\frac{4(3)}{2} = 6.$$

If there are eight treatment conditions, the number of pairwise comparisons is

$$\frac{8(7)}{2} = 28.$$

Do Exercises 3 through 7 at the right.

The Calculation of HSD

The difference between two means is significant at a given α level if it equals or exceeds HSD:

$$\text{HSD} = q_\alpha \sqrt{\frac{s_\varepsilon^2}{n}},$$

in which

s_ε^2 = the error variance,
n = number of subjects in each condition,
q_α = tabled value for a given α-level found in Table E for the error degrees of freedom (df) and the number of means (k).

We shall illustrate the use of the Tukey test by using the worked example appearing in Chapter 8. We obtained an F-ratio of 6.67 which was significant at $\alpha = 0.05$.

The means of the three conditions were as follows:

$$\bar{X}_1 = \frac{\sum X_1}{n_1} = \frac{27}{7} = 3.57,$$

$$\bar{X}_2 = \frac{\sum X_2}{n_2} = \frac{38}{7} = 5.43,$$

$$\bar{X}_3 = \frac{\sum X_3}{n_3} = \frac{56}{7} = 8.00.$$

We also found s_W^2 to be 5.19, with 18 df.

In the following exercises, indicate the number of pairwise comparisons that can be made.

3. $k = 3$

4. $k = 5$

5. $k = 6$

6. $k = 7$

7. $k = 2$

Step 1.   Prepare a matrix showing the mean of each condition and the difference between pairs of means.

	$\bar{X}_1 = 3.57$	$\bar{X}_2 = 5.43$	$\bar{X}_3 = 8.00$
$\bar{X}_1 = 3.57$	—	1.86	4.43
$\bar{X}_2 = 5.43$		—	2.57
$\bar{X}_3 = 8.00$			—

Step 2.   Referring to Table E under error df $= 18$, $k = 3$, and $\alpha = 0.05$, we find $q_\alpha = 3.61$.

Step 3.   Find HSD by multiplying $q_{\alpha = 0.05}$ by

$$\sqrt{\frac{s_W^2}{n}}.$$

In our present example, $s_W^2 = 5.18$, and n per condition is 7. Thus

$$\text{HSD} = 3.61 \sqrt{\frac{5.19}{7}}$$

$$= (3.61)(0.86)$$

$$= 3.10.$$

Step 4.   Referring to the matrix in Step 1, we find that only one pairwise difference between means is significant. We conclude that the mean of condition X_3 is significantly greater than the mean of condition X_1.

Do Exercise 8 at the right.

8.   Apply the HSD test to the analysis of variance of data in Exercise 23, Chapter 8.

CHAPTER 9 TEST

To complete this test, students will require access to Table E (Percentage Points of the Studentized Range).

1. When using the Tukey test to make pairwise comparisons among means, how many comparisons will six treatment conditions yield?

   a) 6   b) 3   c) 30   d) 15

2. Given $\bar{X}_1 = 1.78$, $\bar{X}_2 = 2.45$, $\bar{X}_3 = 3.96$, and HSD $= 1.50$; which means differ significantly from one another?

   a) $\bar{X}_1$ versus $\bar{X}_3$ and $\bar{X}_2$ versus $\bar{X}_3$

   b) $\bar{X}_1$ versus $\bar{X}_2$, $\bar{X}_1$ versus $\bar{X}_3$, and $\bar{X}_2$ versus $\bar{X}_3$

   c) $\bar{X}_2$ versus $\bar{X}_3$

   d) All pairwise comparisons are statistically significant.

3. Given $q_\alpha = 4.02$, $s_W^2 = 27$, $n = 9$; HSD equals:

   a) 20.90   b) 6.95   c) 12.06   d) 3.47

4. Given $\alpha = 0.01$, error df $= 40$, $k = 9$; q_α equals:

   a) 4.63   b) 4.52   c) 5.50   d) 5.39

5. Given $\bar{X}_1 = 2.01$; $\bar{X}_2 = 5.67$; $\bar{X}_3 = 2.68$; $\bar{X}_4 = 4.93$, HSD $= 2.90$; which means differ significantly from one another?

   a) $\bar{X}_1$ versus $\bar{X}_2$, $\bar{X}_1$ versus $\bar{X}_4$, $\bar{X}_2$ versus $\bar{X}_3$ ·

   b) $\bar{X}_1$ versus $\bar{X}_2$, $\bar{X}_1$ versus $\bar{X}_4$

   c) $\bar{X}_1$ versus $\bar{X}_2$, $\bar{X}_2$ versus $\bar{X}_3$

   d) All pairwise comparisons achieve statistical significance.

Practical

6. Summarized below is a one-way analysis of variance of three treatment groups. Conduct a Tukey multicomparison test for pairwise differences among means. Use $\alpha = 0.05$.

Source of variation	Sum of squares	Degrees of freedom	Variance estimate	F
Between groups	35.34	2	17.67	9.30
Within groups	45.53	24	1.90	
Total	80.87	26		

$\sum X_1 = 127.3$

$\sum X_2 = 116.9$

$\sum X_3 = 142.0$

$n_1 = n_2 = n_3 = 9$

One-Way Analysis of Variance, Correlated Samples without Repeated Measures (Randomized Block Design)

Chapter

In Chapter 6 we saw that many factors contribute to the variability of scores in behavioral research. Among the most important are factors like individual differences, which go unidentified and unquantified in an independent samples design. Consequently, the error term is inflated, i.e., it is larger than it would be if such sources of variability were identified, quantified, and "removed" from error. A correlated samples design represents one method for removing important sources of variability from error and thereby providing a more sensitive basis for evaluating differences among means.

THE MAIN CLASSES OF CORRELATED SAMPLES DESIGN

As we saw in Chapter 6, there are two main classes of correlated samples design—repeated measures (before-after) and matched group designs. An analysis of variance is appropriate when $k \geq 3$.

In a repeated measures design, each subject may participate in several or all conditions. The most common type of repeated measures design involves learning tasks in which "trials" constitute the treatment levels.

In a matched group design, subjects are formed into matched groups or blocks according to their similarity on some measure either presumed or known to be correlated with the dependent measure.

Example

In a five-condition experiment using a matched group design, blocks of five subjects are formed on the basis of their similarity to one another on some measure thought to correlate with the dependent measure. The subjects are then randomly assigned to the experimental treatments.

Imagine that 10 subjects received the following scores on a matching variable.

Subject	Score
A	15
B	14
C	10
D	15
E	13
F	11
G	16
H	8
I	9
J	10

OBJECTIVES

1. Know the two main classes of correlated samples designs for which analysis of variance techniques are appropriate.

2. Know how to calculate the three sums of squares that contribute to the total sum of squares.

3. Know how to calculate the F-ratio and to evaluate the outcome of the experiment.

10

Subjects A, B, D, E, and G would form one block. Subjects C, F, H, I, and J would form a second block.

The S's are then assigned at random to the treatments within a block. One possible outcome of the assignment could be as follows.

Block	**Treatment condition**				
	1	**2**	**3**	**4**	**5**
1	B	D	A	E	G
2	F	J	H	I	C

In this chapter, we shall consider only the matched group design.

Do Exercises 1 and 2 at the right.

PARTITIONING THE SUM OF SQUARES

The total sum of squares $\left(\sum x_T^2\right)$ may be partitioned into three components: the between-group sum of squares $\left(\sum x_B^2\right)$, the blocks sum of squares $\left(\sum x_{bl}^2\right)$, and a residual sum of squares representing the interaction of blocks and the between-group (treatment) variable $\left(\sum x_{bl \times B}^2\right)$.

The between-group sum of squares represents the treatment effects; the blocks sum of squares represents the blocking or matching variable, and the residual sum of squares represents random error.

Example

Given the following data in which there are three treatment conditions and twenty-one subjects formed into seven blocks. Determine the significance of the treatment effects, employing $\alpha = 0.01$.

Block	**Treatment condition**			Block sum
	X_1	X_2	X_3	
1	15	13	11	39
2	13	9	10	32
3	12	10	9	31
4	11	13	12	36
5	9	5	7	21
6	8	6	4	18
7	7	5	2	14
	$\sum X_1 = 75$	$\sum X_2 = 61$	$\sum X_3 = 55$	

Step 1. Find $\sum X$ for each condition. Add these together to find $\sum X_T$. In the present problem,

$$\sum X_T = \sum X_1 + \sum X_2 + \sum X_3$$

$$= 75 + 61 + 55$$

$$= 191.$$

Do Exercise 3 at the right.

Step 2. Square $\sum X_T$ and divide by N. With the present data,

$$\frac{(\sum X_T)^2}{N} = \frac{(191)^2}{21} = 1737.19.$$

Do Exercise 4 at the right.

Step 3. Square each score in each condition and sum to obtain $\sum X^2$ for each group. Add these together to obtain $\sum X_T^2$. In the present problem,

$$\sum X_T^2 = \sum X_1^2 + \sum X_2^2 + \sum X_3^2$$

$$= 853 + 605 + 515$$

$$= 1973.$$

Do Exercise 5 at the right.

Step 4. Find the total sum of squares by subtracting the value found in Step 2 from the sum found in Step 3:

$$\sum x_T^2 = \sum X_T^2 - \frac{(\sum X_T)^2}{N}$$

$$= 1973 - 1737.19$$

$$= 235.81.$$

Do Exercise 6 at the right.

Step 5. Find the total number of degrees of freedom:

$$df_T = N - 1$$

$$= 21 - 1$$

$$= 20.$$

Do Exercise 7 at the right.

Given the following data in which there are four treatment conditions and twenty subjects formed into five blocks. Use $\alpha = 0.05$ in the evaluation of the effectiveness of treatment effects.

Block	X_1	X_2	X_3	X_4
1	16	18	18	20
2	14	17	19	26
3	12	13	18	17
4	10	14	13	15
5	8	7	9	11

3. Find $\sum X_T$.

4. Find $(\sum X_T)^2$ from Exercise 3.

5. Find $\sum X_T^2$ for the above data.

6. Find $\sum x_T^2$ for the preceding data.

7. Find df_T for the preceding data.

10

Step 6. Place the values found in Steps 4 and 5 in the summary table found in Step 16.

Do Exercise 8 at the right.

Step 7. Find the between-group sum of squares by squaring each $\sum X$, dividing by the N in the condition, and summing. Subtract the value found in Step 2 from this sum. In the present problem,

$$\sum x_B^2 = \frac{(\sum X_1)^2}{N_1} + \frac{(\sum X_2)^2}{N_2} + \frac{(\sum X_3)^2}{N_3} - 1737.19$$

$$= \frac{(75)^2}{N_1} + \frac{(61)^2}{N_2} + \frac{(55)^2}{N_3} - 1737.19.$$

Since the N's are equal, there will be less rounding error if we put all the quantities to be squared over the least common denominator (7):

$$\sum x_B^2 = \frac{(75)^2 + (61)^2 + (55)^2}{7} - 1737.19$$

$$= 1767.29 - 1737.19$$

$$= 30.10.$$

Do Exercise 9 at the right.

Step 8. Find $df_B = k - 1$. In the present problem, the number of groups or treatment conditions (k) is 3. Thus $df_B = 3 - 1 = 2$.

Do Exercise 10 on the right.

Step 9. Place the values found in Steps 7 and 8 in the summary table found in Step 16.

Do Exercise 11 at the right.

Step 10. Find the blocks sum of squares. This can be found by summing the scores in each block, squaring, dividing by the n in each block, and summing the squared values. Then

$$\frac{(\sum X_T)^2}{N}$$

is subtracted from this quantity:

$$\sum x_{bl}^2 = \frac{(\sum X_{bl_1})^2}{n_{bl_1}} + \frac{(\sum X_{bl_2})^2}{n_{bl_2}} + \cdots + \frac{\sum X_{bl_k}}{n_{bl_k}} - \frac{(\sum X_T)^2}{N},$$

8. Place the values found in Exercises 6 and 7 in the summary table found in Exercise 18.

9. Find $\sum x_B^2$ for the preceding data.

10. Find df_B for the above data.

11. Place the values found in Exercises 9 and 10 in the summary table found in Exercise 18.

in which $\sum x_{bl_k}$ is the sum of the last block and n_{bl_k} is the n in this block.

However, since the n's are equal, it's preferable to employ the following raw score formula:

$$\sum x_{bl}^2 = \frac{(\sum X_{bl_1})^2 + (\sum X_{bl_2})^2 + \cdots + (\sum X_{bl_k})^2}{n_{bl}} - \frac{(\sum X_T)^2}{N},$$

in which n_{bl} is the n in any block.

In the present problem,

$$\sum x_{bl}^2 = \frac{(39)^2 + (32)^2 + (31)^2 + (36)^2 + (21)^2 + (18)^2 + (14)^2}{3}$$

$$- 1737.19$$

$$= \frac{5763}{3} - 1737.19$$

$$= 1921 - 1737.19$$

$$= 183.81.$$

Do Exercise 12 at the right.

Step 11.   Find the number of degrees of freedom for blocks:

$$df_{bl} = bl - 1,$$

in which bl is the number of blocks.   In the present problem,

$$df = bl - 1$$

$$= 7 - 1$$

$$= 6.$$

Do Exercise 13 at the right.

Step 12.   Place the values found in Steps 10 and 11 in the summary table found in Step 16.

Do Exercise 14 at the right.

Step 13.   Find the residual (block × treatments interaction) sum of squares by subtraction:

$$\sum x_{bl \times B}^2 = \sum x_T^2 - (\sum x_B^2 + \sum x_{bl}^2)$$

$$= 235.81 - (30.10 + 183.81)$$

$$= 21.90.$$

12. Find the blocks sum of squares for the preceding data.

13. Find the number of degrees of freedom for blocks in the preceding data.

14. Enter the values found in Exercises 12 and 13 in the summary table in Exercise 18.

10

Do Exercise 15 at the right.

Step 14. Find the number of degrees of freedom for the residual term:

$$df_{bl \times B} = (bl - 1)(B - 1)$$
$$= (6)(2)$$
$$= 12.$$

Do Exercise 16 at the right.

Step 15. Place the values found in Steps 12 and 13 in the summary table found in Step 16.

Do Exercise 17 at the right.

Step 16. Summary table for representing the relevant statistics in a randomized block design:

Source of variation	Sum of squares	Degrees of freedom	Variance estimate	F
Between groups	30.10	2	15.05	
Between blocks	183.81	6	—	
Residual (error)	21.90	12	1.82	
Total	235.81	20		

Step 17. Check to ascertain that

$$\sum x_T^2 = \sum x_B^2 + \sum x_{bl}^2 + \sum x_{bl \times B}^2$$

and

$$df_T = df_B + df_{bl} = df_{bl \times B}.$$

Do Exercise 19 at the right.

Step 18. Find the between-group variance estimate by dividing the between-group sum of squares by df_B. In the present problem,

$$s_B^2 = \frac{30.10}{2} = 15.05.$$

Place in the summary table shown in Step 16.

Do Exercise 20 at the right.

15. Find the residual sum of squares for the preceding data.

16. Find the number of degrees of freedom for the residual term in the preceding problem.

17. Enter the values found in Exercises 15 and 16 in the summary table found in Exercise 18.

18.

Source of variation	Sum of squares	Degrees of freedom	Variance estimate	F
Between groups		3		
Between blocks		4		
Residual (error)		12		
Total		20		

19. Check both the sum of squares and the df in the summary table.

20. Find the between-group variance estimate for the data summarized in Exercise 18.

Step 19. Find the residual variance (error) estimate by dividing the blocks × between-groups sum of squares by $df_{bl \times B}$:

$$s^2_{bl \times B} = \frac{21.90}{12} = 1.82.$$

Place in the summary table shown in Step 16.

Do Exercise 21 at the right.

Step 20. Find the *F*-ratio by the following:

$$F = \frac{s^2_B}{s^2_{bl \times B}} = \frac{15.05}{1.82}$$

$$= 8.27, \qquad df\ 2/12.$$

Do Exercise 22 at the right.

Step 21. Look up $F = 8.27$ in Table C, under 2 and 12 df. If $F = 8.27$ equals or exceeds the tabled value at $\alpha = 0.01$ (boldface type), we reject H_0 and assert that the samples were not drawn from a common population of means.

Since $F = 8.27 > 6.93$, we reject H_0. Apparently the experimental conditions have produced an effect.

Do Exercise 23 at the right.

[*Note:* Since the overall *F*-ratio is statistically significant, you should use the Tukey HSD test (Chapter 9) to evaluate pairwise differences among means. The residual variance estimate, $s^2_{bl \times B}$, replaces s^2_ε in the formula for HSD.]

21. Find the residual variance (error) estimate for the data summarized in Exercise 18.

22. Find the *F*-ratio from the data summarized in Exercise 18.

23. Determine the significance of the obtained *F* at $\alpha = 0.05$ and draw the appropriate conclusion.

CHAPTER 10 TEST

1. In a matched group design, 9 subjects were formed into three blocks on the basis of a preliminary test. The scores were as follows: A, 10; B, 15; C, 36; D, 19; E, 28; F, 5; G, 30; H, 17; I, 8. One block would consist of the following subjects:

    a) C, E, H b) B, D, H c) A, B, F d) C, H, I

2. In a randomized block design:

    a) subjects are assigned at random to experimental conditions

    b) subjects are assigned at random to blocks

    c) subjects are assigned at random to both experimental conditions and blocks

    d) subjects are assigned at random to experimental treatments within blocks

3. Given $\sum X_{bl_1} = 15$, $\sum X_{bl_2} = 26$, $\sum X_{bl_3} = 30$, $k = 5$; $(\sum X_T)^2/N$ equals:

    a) 336.07   b) 360.20   c) 1008.20   d) 120.07

4. Given $N = 45$, $k = 5$, bl $= 9$; df_T equals:

    a) 44   b) 8   c) 36   d) 4

5. Given $N = 90$, $k = 3$, bl $= 30$; df_{bl} equals:

    a) 58   b) 89   c) 87   d) 29

6. Given $N = 60$, $k = 4$, bl $= 15$; $df_{bl \times B}$ equals:

    a) 14   b) 59   c) 42   d) 60

7. Given $\sum x_T^2 = 405$, $\sum x_{bl \times B}^2 = 105$, $\sum x_B^2 = 150$; $\sum x_{bl}^2$ equals:

    a) 660   b) 300   c) 255   d) 150

8. Given $\sum X_{bl_1} = 12$, $\sum X_{bl_2} = 22$, $\sum X_{bl_3} = 25$, number of blocks equals 4; $(\sum X_T)^2/N$ equals:

    a) 813.25   b) 104.42   c) 290.08   d) 870.25

9. Given $\sum x_T^2 = 180$, $df_T = 27$, $\sum x_B^2 = 40$, df $= 6$, $\sum x_{bl \times B}^2 = 80$, df $= 18$; $\sum x_{bl}^2$ equals:

    a) 60   b) 30   c) 300   d) 4.44

10. Given the same data as in Question 9; s_B^2 equals:

    a) 40   b) 6.67   c) 4.44   d) 1.50

11. Given the same data as in Question 9; F equals:

    a) 0.50   b) 9.01   c) 4.50   d) 1.50

12. Given $s_B^2 = 68.72$, df $= 5$, and $s_{bl \times B}^2 = 16.01$, df $= 20$, $\alpha = 0.01$; we conclude:

   a) $F = 4.29$, reject H_0 b) $F = 17.18$, reject H_0

   c) $F = 4.29$, fail to reject H_0 d) $F = 17.18$, fail to reject H_0.

Practical

13. The following scores were obtained by 16 subjects on a preliminary test. Form these scores into four blocks of four subjects each:
   A, 10; B, 25; C, 15; D, 40; E, 11; F, 18; G, 12; H, 17; I, 35; J, 33; K, 19; L, 37; M, 8; N, 23; O, 29; P, 16.

14. The following scores were obtained in a study employing a randomized block design with four treatment groups and six blocks. Analyze the results using $\alpha = 0.01$.

Block	Treatment condition		
	X_1	X_2	X_3
1	26	18	23
2	22	14	21
3	18	19	20
4	14	12	16
5	10	6	11
6	8	7	9

Two-Way Analysis of Variance, Factorial Design Employing Independent Samples

Among the reasons for the use of analysis-of-variance techniques in research are: (1) it makes possible the simultaneous assessment of several levels of a treatment or experimental variable, thereby freeing the researcher from the limitations of the traditional two-group experimental/control design; (2) it permits the evaluation of more than one variable at a time, and makes possible the assessment of possible interactions between and among variables; (3) it represents an efficient means of using research time and effort since in many analysis of variance designs every observation provides information about each variable, interaction of variables, and error.

In a two-way analysis of variance, there are two experimental or treatment variables. One or both of these variables may be either qualitative or quantitative. (See Chapter 8 for the distinction between qualitative and quantitative variables.)

While there are only two experimental variables in a two-way analysis of variance, there may be any number of subclasses or levels of treatment of each variable. A given study might involve two levels of one variable and four levels of a second variable, or three levels of two (each) variables, etc. The traditional way of designating a two-way analysis of variance is by citing the number of levels (or subclasses) of each variable.

Examples

a) A study with two levels of one variable and four levels of a second variable is referred to as having a 2 × 4 design.

b) A study with three levels of each variable is referred to as having a 3 × 3 design.

c) A study with three levels of one variable and four levels of a second variable is referred to as having a 3 × 4 design.

Do Exercises 1 through 5 at the right.

THE CONCEPT OF A TREATMENT COMBINATION

In a factorial design, some level of each treatment variable is administered to each experimental subject. The particular combination of experimental conditions is referred to as a treatment combination. For example, if, in a 3 × 4 factorial design, a given subject is administered the second level of variable A and the third level of variable B, the subject's treatment combination is A_2B_3.

Examples

a) Treatment combinations in a 2 × 2 factorial design

OBJECTIVES

1. Know the terminology used to designate different types of two-way analyses of variance.

2. Know how to partition the total sum of squares into its four components.

3. Know how to apply and interpret the test of significance.

EXERCISES

Show how to designate each of the following designs.

1. Variable A, 4 levels; Variable B, 2 levels.

2. Variable A, 3 levels; Variable B, 5 levels.

3. Variable A, 2 levels; Variable B, 2 levels.

4. Variable A, 2 levels; Variable B, 6 levels.

5. Variable A, 5 levels; Variable B, 7 levels.

11

	A_1		A_2	
	B_1	B_2	B_1	B_2
Treatment combination	A_1B_1	A_1B_2	A_2B_1	A_2B_2

b) Treatment combinations in a 2 × 3 factorial design

	A_1			A_2		
	B_1	B_2	B_3	B_1	B_2	B_3
Treatment combination	A_1B_1	A_1B_2	A_1B_3	A_2B_1	A_2B_2	A_2B_3

Do Exercises 6 through 8 at the right.

The total number of treatment combinations in any factorial design is equal to the product of the treatment levels of all factors or variables.

Examples

In a 2 × 2 factorial design, there are four treatment combinations.

In a 2 × 3 factorial design, there are six treatment combinations.

In more complex factorial designs, the same principle applies. In a 2 × 3 × 4 factorial design, there are 24 treatment combinations.

Do Exercises 9 through 13 at the right.

PARTITIONING THE SUM OF SQUARES

In this chapter, we shall look at the analysis of a 3 × 3 factorial design. However, the analysis may be readily generalized to any number of levels or subclasses of each of the two variables.

In a two-way analysis of variance, the total sum of squares is partitioned into two broad components—within-group sum of squares and treatment combinations sum of squares:

$$\sum x_T^2 = \sum x_W^2 + \sum x_{TC}^2.$$

The treatment combinations sum of squares is itself partitioned into three components—sum of squares for the A variable ($\sum x_A^2$), sum of squares for the B variable ($\sum x_B^2$), and sum of squares for the interaction of the A and B variables ($\sum x_{A \times B}^2$):

$$\sum x_{TC}^2 = \sum x_A^2 + \sum x_B^2 + \sum x_{A \times B}^2.$$

6. Show the treatment combinations in a 4 × 2 factorial design.

7. Show the treatment combinations in a 3 × 3 factorial design.

8. Show the treatment combinations in a 5 × 3 factorial design.

Indicate the number of treatment combinations in the following factorial designs.

9. 2 × 6

10. 3 × 3

11. 3 × 4

12. 4 × 5

13. 2 × 5 × 6

Do Exercises 14 and 15 at the right.

When divided by the appropriate number of degrees of freedom, each of the treatment effects provides an independent estimate of the population variances:

$$\text{est } \sigma_A^2 = \frac{\sum x_A^2}{\text{df}_A}, \qquad \text{df}_A = A - 1,$$

in which A is the number of levels of A;

$$\text{est } \sigma_B^2 = \frac{\sum x_B^2}{\text{df}_B}, \qquad \text{df}_B = B - 1,$$

in which B is the number of levels of B;

$$\text{est } \sigma_{A \times B}^2 = \frac{\sum x_{A \times B}^2}{\text{df}_{A \times B}}, \qquad \text{df}_{A \times B} = \text{df}_A \times \text{df}_B.$$

When divided by the appropriate degrees of freedom, the within-group sum of squares provides an independent estimate of the population variance:

$$\text{est } \sigma_W^2 \text{ (error)} = \frac{\sum x_W^2}{\text{df}_N}, \qquad \text{df}_N = N - TC,$$

in which TC is the number of treatment combinations.

Do Exercises 16 and 17 at the right.

A Worked Example Using a 3 × 3 Factorial Design

	A_1			A_2			A_3		
	B_1	B_2	B_3	B_1	B_2	B_3	B_1	B_2	B_3
	2	3	6	1	2	5	4	5	6
	4	5	7	3	5	8	5	8	10
	5	8	9	6	7	6	3	7	12
	7	11	12	7	9	11	7	9	11
Treatment combination	A_1B_1	A_1B_2	A_1B_3	A_2B_1	A_2B_2	A_2B_3	A_3B_1	A_3B_2	A_3B_3
$\sum$	18	27	34	17	23	30	19	29	39

$$\sum X_T = 236$$

Determine if there is a significant effect of the A variable, the B variable, or an interaction between the two variables. Use $\alpha = 0.01$.

14. Show the partitioning of the total sum of squares into two broad components in a two-variable design. The two variables are designated C and D.

15. Referring back to Exercise 14, show the partitioning of the treatment combinations sum of squares into its three components.

16. Referring back to Exercise 14, show how to obtain variance estimates from the treatment groups and their interaction.

17. Referring back to Exercise 14, show how to obtain the within-group variance or error estimate.

Step 1. Find $\sum x_T^2$. Square each score, sum all of the squared values, and subtract $(\sum X_T)^2/N$:

$$\sum x_T^2 = \sum X_T^2 - \frac{(\sum X_T)^2}{N}.$$

In the present example,

$$\sum x_T^2 = 2^2 + 4^2 + 5^2 + \cdots + 11^2 - \frac{(236)^2}{36}$$

$$= 1842 - 1547.11$$

$$= 294.89.$$

Step 2. Find df_T:

$$df_T = N - 1$$

$$= 36 - 1$$

$$= 35.$$

Step 3. Enter $\sum x_T^2$ and df_T into the summary table found in Step 19.

Do Exercise 18 at the right.

Step 4. Find the sum of squares for the treatment combinations. In the present problem,

$$\sum x_{TC} = \frac{(\sum A_1 B_1)^2}{n_{A_1 B_1}} + \frac{(\sum A_1 B_2)^2}{n_{A_1 B_2}} + \cdots + \frac{(\sum A_3 B_3)^2}{n_{A_3 B_3}} - \frac{(\sum X_T)^2}{N}$$

$$= \frac{(18)^2}{4} + \frac{(27)^2}{4} + \cdots + \frac{(39)^2}{4} - 1547.11.$$

Since n is the same for each treatment combination, there will be less rounding error if the squares of the treatment combinations are summed and then divided by the n in each group:

$$\sum x_{TC}^2 = \frac{(\sum A_1 B_1)^2 + (\sum A_1 B_2)^2 + \cdots + (\sum A_3 B_3)^2}{n_{TC}} - \frac{(\sum X_T)^2}{N}$$

$$= \frac{(18)^2 + (27)^2 + (34)^2 + (17)^2 + (23)^2 + (30)^2 + (19)^2 + (29)^2 + (39)^2}{4}$$

$$- 1547.11$$

$$= \frac{6650}{4} - 1547.11$$

$$= 1662.50 - 1547.11$$

$$= 115.39.$$

Refer to the following data on a 2×3 factorial design for Exercises 18 through 33. Use $\alpha = 0.05$.

A_1			A_2		
B_1	B_2	B_3	B_1	B_2	B_3
4	5	4	8	6	11
6	8	6	7	9	9
7	3	2	9	12	13
9	7	5	11	8	7

18. Find $\sum x_T^2$ and df_T. Enter into the summary table shown in Exercise 24.

Step 5. Find the number of degrees of freedom for the treatment combinations sum of squares:

$$\mathrm{df}_{TC} = TC - 1,$$

in which TC is the number of treatment combinations. In the present example,

$$\mathrm{df}_{TC} = 9 - 1 = 8.$$

Step 6. Enter the sum of squares of treatment combinations and the corresponding degrees of freedom in the summary table found in Step 19.

Do Exercise 19 at the right.

Step 7. Begin the partitioning of the treatment combinations sum of squares by finding the sum of squares for the A variable:

$$\sum x_A^2 = \frac{\left(\sum X_{A_1}\right)^2}{n_{A_1}} + \frac{\left(\sum X_{A_2}\right)^2}{n_{A_2}} + \frac{\left(\sum X_{A_3}\right)^2}{n_{A_3}} - \frac{\left(\sum X_T\right)^2}{N}$$

$$= \frac{(79)^2}{12} + \frac{(70)^2}{12} + \frac{(87)^2}{12} - 1547.11.$$

Since n is the same for each A condition, there will be less rounding error if the squares of each level of A are summed and then divided by the n in each A condition:

$$\sum x_A^2 = \frac{\left(\sum X_{A_1}\right)^2 + \left(\sum X_{A_2}\right)^2 + \left(\sum X_{A_3}\right)^2}{n_A} - \frac{\left(\sum X_T\right)^2}{N}$$

$$= \frac{(79)^2 + (70)^2 + (87)^2}{n_A} - 1547.11$$

$$= \frac{18710}{12} - 1547.11$$

$$= 1559.17 - 1547.11$$

$$= 12.06.$$

Step 8. Find the number of degrees of freedom for the A condition:

$$\mathrm{df}_A = A - 1,$$

in which A is the number of levels of A. In the present example,

$$\mathrm{df}_A = 3 - 1$$

$$= 2.$$

19. Find the sum of squares and df for the treatment combinations. Enter into the summary table shown in Exercise 24.

11

Step 9. Enter the sum of squares for the A variable and the corresponding degrees of freedom in the summary table found in Step 19.

Do Exercise 20 at the right.

Step 10. Find the sum of squares of the B variable.

$$\sum x_B^2 = \frac{(\sum X_{B_1})^2}{n_{B_1}} + \frac{(\sum X_{B_2})^2}{n_{B_2}} + \frac{(\sum X_{B_3})^2}{n_{B_3}} - \frac{(\sum X_T)^2}{N}$$

$$= \frac{(54)^2}{12} + \frac{(79)^2}{12} + \frac{(103)^2}{12} - 1547.11.$$

Since n is the same for each B condition, there will be less rounding error if the squares of each level of B are summed and then divided by the n in each condition:

$$\sum x_B^2 = \frac{(54)^2 + (79)^2 + (103)^2}{12} - 1547.11$$

$$= \frac{19,766}{12} - 1547.11$$

$$= 1647.17 - 1547.11$$

$$= 100.06.$$

Step 11. Find the number of degrees of freedom for the B condition:

$$\text{df}_B = B - 1,$$

in which B is the number of levels of B. In the present example,

$$\text{df}_B = 3 - 1$$

$$= 2.$$

Step 12. Enter the sum of squares for the B variable and the corresponding degrees of freedom in the summary table found in Step 19.

Do Exercise 21 at the right.

Step 13. Find the interaction sum of squares by subtraction:

$$\sum x_{A \times B}^2 = \sum x_{TC}^2 - \left(\sum x_A^2 + \sum x_B^2\right)$$

$$= 115.39 - (12.06 + 100.06)$$

$$= 3.27.$$

20. Find the sum of squares and df for the A variable. Enter into the summary table shown in Exercise 24.

21. Find the sum of squares and df for the B variable. Enter into the summary table shown in Exercise 24.

Step 14. Find the number of degrees of freedom for the interaction sum of squares:

$$df_{A \times B} = (A - 1)(B - 1)$$
$$= (2)(2)$$
$$= 4.$$

Step 15. Enter the interaction sum of squares and the corresponding degrees of freedom in the summary table found in Step 19.

Do Exercise 22 at the right.

Step 16. Find the within-group sum of squares. The within-group sum of squares may be obtained by subtraction:

$$\sum x_W^2 = \sum x_T^2 - \sum x_{TC}^2$$
$$= 294.89 - 115.39$$
$$= 179.5.$$

It may also be obtained directly using the following formula:

$$\sum x_W^2 = \left[\sum X_{TC_1}^2 - \frac{(\sum X_{TC_1})^2}{n_{TC_1}} \right] + \left[\sum X_{TC_2}^2 - \frac{(\sum X_{TC_2})^2}{n_{TC_2}} \right] + \cdots$$
$$+ \left[\sum X_{TC_k}^2 - \frac{(\sum X_{TC_k}^2)^2}{n_{TC_k}} \right],$$

in which TC_1 is the first treatment combination and TC_k the last.

In the present example;

$$\sum x_W^2 = \left[94 - \frac{(18)^2}{4} \right] + \left[219 - \frac{(27)^2}{4} \right] + \left[310 - \frac{(34)^2}{4} \right]$$
$$+ \left[95 - \frac{(17)^2}{4} \right] + \left[159 - \frac{(23)^2}{4} \right] + \left[246 - \frac{(30)^2}{4} \right]$$
$$+ \left[99 - \frac{(19)^2}{4} \right] + \left[219 - \frac{(29)^2}{4} \right] + \left[401 - \frac{(39)^2}{4} \right]$$
$$= 94 + 219 + 310 + 95 + 159 + 246 + 99 + 219 + 401$$
$$- \frac{(18)^2 + (27)^2 + (34)^2 + (17)^2 + (23)^2 + (30)^2 + (19)^2 + (29)^2 + (39)^2}{4}$$
$$= 1842 - \frac{6650}{4}$$
$$= 1842 - 1662.5*$$
$$= 179.5.$$

22. Find the interaction sum of squares and the corresponding df. Enter into the summary table shown in Exercise 24.

This agrees with the result we obtained by direct subtraction. Note, in the expression indicated by the asterisk, that the first item on the right of the equality sign is $\sum X_T^2$ and the second term is $\sum X_{TC}^2$.

Step 17. Find the within-group degrees of freedom:

$$\text{df}_W = N - TC$$
$$= 36 - 9$$
$$= 27.$$

Step 18. Enter the within-group sum of squares and degrees of freedom in the summary table shown in Step 19.

Do Exercise 23 at the right.

Step 19. Summary table

Source of variation	Sum of squares	Degrees of freedom	Variance estimate	*F*
Treatment combinations		115.39	8		
A variable	12.06	2	6.03	0.91
B variable	100.06	2	50.03	7.52
A × *B*	3.27	4	0.82	0.12
Within-group (error)		179.50	27	6.65	
Total		294.89	35		

Step 20. Check to ascertain that

$$\sum x_{TC}^2 + \sum x_W^2 = \sum x_T^2$$

and

$$\sum x_A^2 + \sum x_B^2 + \sum x_{A \times B}^2 = \sum x_{TC}^2.$$

Do Exercise 25 at the right.

Step 21. Find the $A \times B$ interaction variance estimate by dividing the interaction sum of squares by $\text{df}_{A \times B}$:

$$s_{A \times B}^2 = \frac{\sum x_{A \times B}^2}{\text{df}_{A \times B}}$$
$$= \frac{3.27}{4}$$
$$= 0.82.$$

23. Find the within-group sum of squares and degrees of freedom. Enter into the summary table shown in Exercise 24.

24. Summary table—analysis of a 2 × 3 factorial design

Source of variation	Sum of squares	Degrees of freedom	Variance estimate	*F*
Treatment combinations				
A variable				
B variable				
A × *B*				
Within-group (error)				
Total				

25. Check to ascertain that all values in the summary table are reconciled.

Place in the summary table shown in Step 19.

Do Exercise 26 at the right.

Step 22.   Find the B variable variance estimate by dividing the B variable sum of squares by df_B:

$$s_B^2 = \frac{\sum x_B^2}{\mathrm{df}_B}$$

$$= \frac{100.06}{2}$$

$$= 50.03.$$

Do Exercise 27 at the right.

Step 23.   Find the A variable variance estimate by dividing the A variable sum of squares by df_A:

$$s_A^2 = \frac{\sum x_A^2}{\mathrm{df}_A}$$

$$= \frac{12.06}{2}$$

$$= 6.03.$$

Do Exercise 28 at the right.

Step 24.   Find the within-group variance estimate (error) by dividing the within-group sum of squares by df_W.

$$s_W^2 = \frac{\sum x_W^2}{\mathrm{df}_W}$$

$$= \frac{179.50}{27}$$

$$= 6.65.$$

Do Exercise 29 at the right.

Step 25.   Find the interaction F-ratio by dividing the interaction estimated variance by the within-group estimated variance:

$$F = \frac{s_{A \times B}^2}{s_W^2}$$

$$= \frac{0.82}{6.65} = 0.12, \qquad \mathrm{df} = 4/27.$$

26. Find the $A \times B$ interaction variance estimate and enter into the table shown in Exercise 24.

27. Find the B variable variance estimate and enter into the table shown in Exercise 24.

28. Find the A variable variance estimate and enter into the table shown in Exercise 24.

29. Find the within-group variance estimate (error) and enter into the table shown in Exercise 24.

Step 26. Consult Table C under 4 and 27 df to find the critical value of F required to reject H_0 at the $\alpha = 0.01$ level. Since $F_{0.01} \geq 4.11$, we fail to reject H_0.

Do Exercise 30 at the right.

Step 27. Find the B variable F-ratio by dividing the estimated B variance by the within-group estimated variance:

$$F = \frac{s_B^2}{s_W^2}$$

$$= \frac{50.03}{6.65}$$

$$= 7.52, \quad df = 2/27.$$

Step 28. Consult Table C under 2 and 27 df to find the critical value of F required to reject H_0 at the $\alpha = 0.01$ level. Since $F_{0.01} \geq 5.49$, the obtained ratio of 7.52 is in the critical region for rejecting H_0. There is a significant effect of the B variable.

Do Exercise 31 at the right.

Step 29. Find the A variable F-ratio by dividing the estimated A variance by the within-group estimated variance:

$$F = \frac{s_A^2}{s_W^2}$$

$$= \frac{6.03}{6.65}$$

$$= 0.91, \quad df = 2/27.$$

Step 30. Consult Table C under 2 and 27 df to find the critical value required to reject H_0 at the $\alpha = 0.01$ level. Since $F_{0.01} \geq 5.49$, the obtained F of 0.91 is not within the critical region. We fail to reject H_0.

Do Exercise 32 at the right.

Conclusion: Of the three effects evaluated—A variable, B variable, and the interaction of A and B—only the B variable was found to be statistically significant. It is now appropriate to employ a multicomparison test in order to test pairwise differences among means.

Do Exercise 33 at the right.

30. Find the interaction F-ratio and critical value of F required to reject H_0 at $\alpha = 0.05$. Make the statistical decision whether or not to reject H_0.

31. Find the B variable F-ratio and the critical value of F required to reject H_0 at $\alpha = 0.05$. Make the statistical decision whether or not to reject H_0.

32. Find the A variable F-ratio and the critical value of F required to reject H_0 at $\alpha = 0.05$. Make the statistical decision whether or not to reject H_0.

33. State your conclusions. Conduct any follow-up statistical tests, if warranted.

APPLYING THE TUKEY HSD TEST TO MAKE PAIRWISE COMPARISONS AMONG THE *B*-VARIABLE MEANS

A difference between two means is significant at a given α level if it equals or exceeds HSD:

$$HSD = q_\alpha \sqrt{\frac{s_\varepsilon^2}{n}},$$

in which

s_ε^2 = the error variance estimate (within-group variance in the present example),

n = number of subjects in each condition,

q_α = tabled value for a given α level found in Table E for df and the number of means (k).

We previously found $\sum X_{B_1} = 54$, $\sum X_{B_2} = 79$, and $\sum X_{B_3} = 103$. Since n is 12 in each case, the respective means are

$$\bar{X}_{B_1} = \frac{54}{12} = 4.50,$$

$$\bar{X}_{B_2} = \frac{79}{12} = 6.58,$$

$$\bar{X}_{B_3} = \frac{103}{12} = 8.58.$$

Step 1.   Prepare a matrix showing the mean of each condition and the differences between pairs of means.

	$\bar{X}_{B_1}=$ 4.50	$\bar{X}_{B_2}=$ 6.58	$\bar{X}_{B_3}=$ 8.58
$\bar{X}_{B_1} = 4.50$	—	2.08	4.08
$\bar{X}_{B_2} = 6.58$	—	—	2.00
$\bar{X}_{B_3} = 8.58$	—	—	—

Step 2.   Referring to Table E under error df = 27, $k = 3$, and $\alpha = 0.01$, we find no tabled values. We must interpolate between 24 and 30 df. Since 27 df is halfway between 24 and 30 df, it is halfway between $q = 4.55$ and $q = 4.45$. Therefore, $q_{\alpha=0.01} = 4.50$.

Step 3.   Find HSD by multiplying $q_{\alpha=0.01}$ by

$$\sqrt{\frac{s_W^2}{n}}.$$

The *n* per condition is 12. Thus

$$\text{HSD} = 4.50 \sqrt{\frac{6.65}{12}}$$

$$= (4.50)(0.74)$$

$$= 3.33.$$

Step 4. Referring to the matrix in Step 1, we find that only one pairwise difference between means is significant. We may conclude that $\bar{X}_{B_3}$ is significantly greater than $\bar{X}_{B_1}$.

CHAPTER 11 TEST

1. A 5×6 factorial design involves:

   a) 11 different treatment levels

   b) five independent variables at six treatment levels

   c) five levels of one treatment variable and six levels of another

   d) five dependent and six independent variables.

2. In a 7×8 factorial design, the number of treatment combinations is:

   a) 28   b) 56   c) 42   d) 15

3. If a subject received the fourth level of treatment A and the second level of treatment B in a 5×6 factorial design, the treatment combination would be:

   a) $A_4 B_2$ b) $A_5 B_6$ c) 4×2 d) 5×6

4. In a two-way analysis of variance, the treatment combinations sum of squares may be partioned into the following three components:

   a) total, treatment combinations, within-group

   b) total, within-group, A variable

   c) A-variable, B-variable, within-group

   d) A-variable, B-variable, $A \times B$ interaction

5. In a two-way analysis of variance, each observation yields information concerning how many treatment effects and their interactions?

   a) one   b) two   c) three   d) four

6. In a 3×3 factorial design, the following scores were obtained for each treatment combination: 8, 12, 9, 14, 16, 12, 10, 5, 3. The n in each treatment combination equalled two. $\sum x_{TC}$ equals:

   a) 590.50   b) 113.00   c) 48.22   d) 2325.5

7. In question 6, the sum of each treatment level of the B-variable was 35, 30, and 16. The sum of squares for B was:

   a) 99.94   b) 826.0   c) 32.33   d) 28.74

8. In a 6×7 factorial design in which $N = 210$, df_W equals:

   a) 168   b) 42   c) 209   d) 30

9. In a 4×5 factorial design in which $N = 100$, df_A equals:

   a) 4   b) 99   c) 80   d) 3

10. In a 4×4 factorial design in which $N = 64$, $df_{A \times B}$ equals:

    a) 9   b) 16   c) 63   d) 3

11. In a 3×4 factorial design in which $N = 72$, df_T equals:

    a) 9   b) 71   c) 60   d) 12

12. Given $\sum x_{TC}^2 = 205$, $\sum x_A^2 = 35$, $\sum x_B^2 = 90$, $\sum x_T^2 = 450$; $\sum x_{A \times B}^2$ equals:

    a) 245   b) 80   c) 325   d) 655

13. Given $\sum x_T^2 = 60$, $\sum x_{TC}^2 = 15$, $\sum x_A^2 = 5$, $\sum x_{A \times B}^2 = 3$; $\sum x_B^2$ and $\sum x_W^2$ equal, respectively:

    a) 7, 45   b) 8, 45   c) 7, 75   d) 23, 75

14. Given $\sum x_{TC}^2 = 180$, $\sum x_A^2 = 70$, $\sum x_B^2 = 60$, $\sum x_{A \times B}^2 = 50$, $\sum x_W^2 = 140$, and $df_A = 2$, $df_B = 3$, $df_{A \times B} = 9$, and $df_W = 60$; s_A^2 equals:

    a) 0.50   b) 15.02   c) 35   d) 23.33

15. Given the same data as in question 14; $s_{A \times B}^2$ equals:

    a) 16.67   b) 25   c) 5.56   d) 2.39

Practical

16. Following are scores made by subjects in a 3×3 factorial experiment. Conduct the appropriate analysis of variance and show conclusions warranted by the analysis. Use $\alpha = 0.05$.

	A_1			A_2			A_3	
B_1	B_2	B_3	B_1	B_2	B_3	B_1	B_2	B_3
8	10	12	5	9	16	6	10	17
1	7	9	3	10	11	5	7	9
7	9	11	8	10	12	4	11	12
9	7	5	8	6	4	10	8	6
4	8	15	6	8	10	8	10	12
12	14	15	10	12	13	13	14	14

The author is grateful to the authors and publishers listed below for permission to adapt from the following tables.

Table B Table III of R. A. Fisher and F. Yates, *Statistical Tables for Biological, Agricultural, and Medical Research*. Edinburgh: Oliver and Boyd, Ltd., 1948.

Table C G. W. Snedecor and William G. Cochran, *Statistical Methods*, 6th ed. Ames, Iowa: Iowa State University Press © 1967.

Table C_1 M. Merrington and C. M. Thompson, "Tables of Percentage Points of the Inverted Beta Distribution." *Biometrika*, **33,** 73, 1943.

Table D J. Sandler, A test of the significance of the difference between the means of correlated measures, based on a simplification of Student's *t*. *Brit. J. Psychol.*, **46,** 225–226, 1955.

Table E E. S. Pearson and H. O. Hartley, *Biometrika Tables for Statisticians*, Vol. 1, 2nd ed. New York: Cambridge, 1958.

Table F Q. McNemar, Table B of *Psychological Statistics*. New York: John Wiley and Sons, Inc., 1962.

Appendix

Table A Percent of area under the standard normal curve, percentile rank corresponding to a given z, and height of ordinate at z

z	(A) Percentile rank (percent of area below z)	(B) Percent of area from mean to z	(C) Percent of area beyond z	(D) Height of ordinate at z
−3.70	00.01	49.99	00.01	.0004
−3.60	00.02	49.98	00.02	.0006
−3.50	00.02	49.98	00.02	.0009
−3.40	00.03	49.97	00.03	.0012
−3.30	00.05	49.95	00.05	.0017
−3.25	00.06	49.95	00.06	.0020
−3.24	00.06	49.94	00.06	.0021
−3.23	00.06	49.94	00.06	.0022
−3.22	00.06	49.94	00.06	.0022
−3.21	00.07	49.93	00.07	.0023
−3.20	00.07	49.93	00.07	.0024
−3.19	00.07	49.93	00.07	.0025
−3.18	00.07	49.93	00.07	.0025
−3.17	00.08	49.92	00.08	.0026
−3.16	00.08	49.92	00.08	.0027
−3.15	00.08	49.92	00.08	.0028
−3.14	00.08	49.92	00.08	.0029
−3.13	00.09	49.91	00.09	.0030
−3.12	00.09	49.91	00.09	.0031
−3.11	00.09	49.91	00.09	.0032
−3.10	00.10	49.90	00.10	.0033
−3.09	00.10	49.90	00.10	.0034
−3.08	00.10	49.90	00.10	.0035
−3.07	00.11	49.89	00.11	.0036
−3.06	00.11	49.89	00.11	.0037
−3.05	00.11	49.89	00.11	.0038
−3.04	00.12	49.88	00.12	.0039
−3.03	00.12	49.88	00.12	.0040
−3.02	00.13	49.87	00.13	.0042
−3.01	00.13	49.87	00.13	.0043
−3.00	00.13	49.87	00.13	.0044
−2.99	00.14	49.86	00.14	.0046
−2.98	00.14	49.86	00.14	.0047
−2.97	00.15	49.85	00.15	.0048
−2.96	00.15	49.85	00.15	.0050
−2.95	00.16	49.84	00.16	.0051
−2.94	00.16	49.84	00.16	.0053
−2.93	00.17	49.83	00.17	.0055
−2.92	00.18	49.82	00.18	.0056
−2.91	00.18	49.82	00.18	.0058
−2.90	00.19	49.81	00.19	.0060
−2.89	00.19	49.81	00.19	.0061
−2.88	00.20	49.80	00.20	.0063
−2.87	00.21	49.79	00.21	.0065
−2.86	00.21	49.79	00.21	.0067
−2.85	00.22	49.78	00.22	.0069
−2.84	00.23	49.77	00.23	.0071
−2.83	00.23	49.77	00.23	.0073
−2.82	00.24	49.76	00.24	.0075
−2.81	00.25	49.75	00.25	.0077

Table A   *(Continued)*

z	(A) Percentile rank (percent of area below z)	(B) Percent of area from mean to z	(C) Percent of area beyond z	(D) Height of ordinate at z
−2.80	00.26	49.74	00.26	.0079
−2.79	00.26	49.74	00.26	.0081
−2.78	00.27	49.73	00.27	.0084
−2.77	00.28	49.72	00.28	.0086
−2.76	00.29	49.71	00.29	.0088
−2.75	00.30	49.70	00.30	.0091
−2.74	00.31	49.69	00.31	.0093
−2.73	00.32	49.68	00.32	.0096
−2.72	00.33	49.67	00.33	.0099
−2.71	00.34	49.66	00.34	.0101
−2.70	00.35	49.65	00.35	.0104
−2.69	00.36	49.64	00.36	.0107
−2.68	00.37	49.63	00.37	.0110
−2.67	00.38	49.62	00.38	.0113
−2.66	00.39	49.61	00.39	.0116
−2.65	00.40	49.60	00.40	.0119
−2.64	00.41	49.59	00.41	.0122
−2.63	00.43	49.57	00.43	.0126
−2.62	00.44	49.56	00.44	.0129
−2.61	00.45	49.55	00.45	.0132
−2.60	00.47	49.53	00.47	.0136
−2.59	00.48	49.52	00.48	.0139
−2.58	00.49	49.51	00.49	.0143
−2.57	00.51	49.49	00.51	.0147
−2.56	00.52	49.48	00.52	.0151
−2.55	00.54	49.46	00.54	.0154
−2.54	00.55	49.45	00.55	.0158
−2.53	00.57	49.43	00.57	.0163
−2.52	00.59	49.41	00.59	.0167
−2.51	00.60	49.40	00.60	.0171
−2.50	00.62	49.38	00.62	.0175
−2.49	00.64	49.36	00.64	.0180
−2.48	00.66	49.34	00.66	.0184
−2.47	00.68	49.32	00.68	.0189
−2.46	00.69	49.31	00.69	.0194
−2.45	00.71	49.29	00.71	.0198
−2.44	00.73	49.27	00.73	.0203
−2.43	00.75	49.25	00.75	.0208
−2.42	00.78	49.22	00.78	.0213
−2.41	00.80	49.20	00.80	.0219
−2.40	00.82	49.18	00.82	.0224
−2.39	00.84	49.16	00.84	.0229
−2.38	00.87	49.13	00.87	.0235
−2.37	00.89	49.11	00.89	.0241
−2.36	00.91	49.09	00.91	.0246
−2.35	00.94	49.06	00.94	.0252
−2.34	00.96	49.04	00.96	.0258
−2.33	00.99	49.01	00.99	.0264
−2.32	01.02	48.98	01.02	.0270
−2.31	01.04	48.96	01.04	.0277

Table A *(Continued)*

z	(A) Percentile rank (percent of area below z)	(B) Percent of area from mean to z	(C) Percent of area beyond z	(D) Height of ordinate at z
−2.30	01.07	48.93	01.07	.0283
−2.29	01.10	48.90	01.10	.0290
−2.28	01.13	48.87	01.13	.0297
−2.27	01.16	48.84	01.16	.0303
−2.26	01.19	48.81	01.19	.0310
−2.25	01.22	48.78	01.22	.0317
−2.24	01.25	48.75	01.25	.0325
−2.23	01.29	48.71	01.29	.0332
−2.22	01.32	48.68	01.32	.0339
−2.21	01.36	48.64	01.36	.0347
−2.20	01.39	48.61	01.39	.0355
−2.19	01.43	48.57	01.43	.0363
−2.18	01.46	48.54	01.46	.0371
−2.17	01.50	48.50	01.50	.0379
−2.16	01.54	48.46	01.54	.0387
−2.15	01.58	48.42	01.58	.0396
−2.14	01.62	48.38	01.62	.0404
−2.13	01.66	48.34	01.66	.0413
−2.12	01.70	48.30	01.70	.0422
−2.11	01.74	48.26	01.74	.0431
−2.10	01.79	48.21	01.79	.0440
−2.09	01.83	48.17	01.83	.0449
−2.08	01.88	48.12	01.88	.0459
−2.07	01.92	48.08	01.92	.0468
−2.06	01.97	48.03	01.97	.0478
−2.05	02.02	47.98	02.02	.0488
−2.04	02.07	47.93	02.07	.0498
−2.03	02.12	47.88	02.12	.0508
−2.02	02.17	47.83	02.17	.0519
−2.01	02.22	47.78	02.22	.0529
−2.00	02.28	47.72	02.28	.0540
−1.99	02.33	47.67	02.33	.0551
−1.98	02.39	47.61	02.39	.0562
−1.97	02.44	47.56	02.44	.0573
−1.96	02.50	47.50	02.50	.0584
−1.95	02.56	47.44	02.56	.0596
−1.94	02.62	47.38	02.62	.0608
−1.93	02.68	47.32	02.68	.0620
−1.92	02.74	47.26	02.74	.0632
−1.91	02.81	47.19	02.81	.0644
−1.90	02.87	47.13	02.87	.0656
−1.89	02.94	47.06	02.94	.0669
−1.88	03.01	46.99	03.01	.0681
−1.87	03.07	46.93	03.07	.0694
−1.86	03.14	46.86	03.14	.0707
−1.85	03.22	46.78	03.22	.0721
−1.84	03.29	46.71	03.29	.0734
−1.83	03.36	46.64	03.36	.0748
−1.82	03.44	46.56	03.44	.0761
−1.81	03.51	46.49	03.51	.0775

Table A   (*Continued*)

z	(A) Percentile rank (percent of area below z)	(B) Percent of area from mean to z	(C) Percent of area beyond z	(D) Height of ordinate at z
−1.80	03.59	46.41	03.59	.0790
−1.79	03.67	46.33	03.67	.0804
−1.78	03.75	46.25	03.75	.0818
−1.77	03.84	46.16	03.84	.0833
−1.76	03.92	46.08	03.92	.0848
−1.75	04.01	45.99	04.01	.0863
−1.74	04.09	45.91	04.09	.0878
−1.73	04.18	45.82	04.18	.0893
−1.72	04.27	45.73	04.27	.0909
−1.71	04.36	45.64	04.36	.0925
−1.70	04.46	45.54	04.46	.0940
−1.69	04.55	45.45	04.55	.0957
−1.68	04.65	45.35	04.65	.0973
−1.67	04.75	45.25	04.75	.0989
−1.66	04.85	45.15	04.85	.1006
−1.65	04.95	45.05	04.95	.1023
−1.64	05.05	44.95	05.05	.1040
−1.63	05.16	44.84	05.16	.1057
−1.62	05.26	44.74	05.26	.1074
−1.61	05.37	44.63	05.37	.1092
−1.60	05.48	44.52	05.48	.1109
−1.59	05.59	44.41	05.59	.1127
−1.58	05.71	44.29	05.71	.1145
−1.57	05.82	44.18	05.82	.1163
−1.56	05.94	44.06	05.94	.1182
−1.55	06.06	43.94	06.06	.1200
−1.54	06.18	43.82	06.18	.1219
−1.53	06.30	43.70	06.30	.1238
−1.52	06.43	43.57	06.43	.1257
−1.51	06.55	43.45	06.55	.1276
−1.50	06.68	43.32	06.68	.1295
−1.49	06.81	43.19	06.81	.1315
−1.48	06.94	43.06	06.94	.1334
−1.47	07.08	42.92	07.08	.1354
−1.46	07.21	42.79	07.21	.1374
−1.45	07.35	42.65	07.35	.1394
−1.44	07.49	42.51	07.49	.1415
−1.43	07.64	42.36	07.64	.1435
−1.42	07.78	42.22	07.78	.1456
−1.41	07.93	42.07	07.93	.1476
−1.40	08.08	41.92	08.08	.1497
−1.39	08.23	41.77	08.23	.1518
−1.38	08.38	41.62	08.38	.1539
−1.37	08.53	41.47	08.53	.1561
−1.36	08.69	41.31	08.69	.1582
−1.35	08.85	41.15	08.85	.1604
−1.34	09.01	40.99	09.01	.1626
−1.33	09.18	40.82	09.18	.1647
−1.32	09.34	40.66	09.34	.1669
−1.31	09.51	40.49	09.51	.1691

Table A   (*Continued*)

z	(A) Percentile rank (percent of area below z)	(B) Percent of area from mean to z	(C) Percent of area beyond z	(D) Height of ordinate at z
−1.30	09.68	40.32	09.68	.1714
−1.29	09.85	40.15	09.85	.1736
−1.28	10.03	39.96	10.03	.1758
−1.27	10.20	39.80	10.20	.1781
−1.26	10.38	39.62	10.38	.1804
−1.25	10.56	39.44	10.56	.1826
−1.24	10.75	39.25	10.75	.1849
−1.23	10.93	39.07	10.93	.1872
−1.22	11.12	38.88	11.12	.1895
−1.21	11.31	38.69	11.31	.1919
−1.20	11.51	38.49	11.51	.1942
−1.19	11.70	38.30	11.70	.1965
−1.18	11.90	38.10	11.90	.1989
−1.17	12.10	37.90	12.10	.2012
−1.16	12.30	37.70	12.30	.2036
−1.15	12.51	37.49	12.51	.2059
−1.14	12.71	37.29	12.71	.2083
−1.13	12.92	37.08	12.92	.2107
−1.12	13.14	36.86	13.14	.2131
−1.11	13.35	36.65	13.35	.2155
−1.10	13.57	36.43	13.57	.2179
−1.09	13.79	36.21	13.79	.2203
−1.08	14.01	35.99	14.01	.2227
−1.07	14.23	35.77	14.23	.2251
−1.06	14.46	35.54	14.46	.2275
−1.05	14.69	35.31	14.69	.2299
−1.04	14.92	35.08	14.92	.2323
−1.03	15.15	34.85	15.15	.2347
−1.02	15.39	34.61	15.39	.2371
−1.01	15.62	34.38	15.62	.2396
−1.00	15.87	34.13	15.87	.2420
−0.99	16.11	33.89	16.11	.2444
−0.98	16.35	33.65	16.35	.2468
−0.97	16.60	33.40	16.60	.2492
−0.96	16.85	33.15	16.85	.2516
−0.95	17.11	32.89	17.11	.2541
−0.94	17.36	32.64	17.36	.2565
−0.93	17.62	32.38	17.62	.2589
−0.92	17.88	32.12	17.88	.2613
−0.91	18.14	31.86	18.14	.2637
−0.90	18.41	31.59	18.41	.2661
−0.89	18.67	31.33	18.67	.2685
−0.88	18.94	31.06	18.94	.2709
−0.87	19.22	30.78	19.22	.2732
−0.86	19.49	30.51	19.49	.2756
−0.85	19.77	30.23	19.77	.2780
−0.84	20.05	29.95	20.05	.2803
−0.83	20.33	29.67	20.33	.2827
−0.82	20.61	29.39	20.61	.2850
−0.81	20.90	29.10	20.90	.2874

Table A   (*Continued*)

z	(A) Percentile rank (percent of area below z)	(B) Percent of area from mean to z	(C) Percent of area beyond z	(D) Height of ordinate at z
−0.80	21.19	28.81	21.19	.2897
−0.79	21.48	28.52	21.48	.2920
−0.78	21.77	28.23	21.77	.2943
−0.77	22.06	27.94	22.06	.2966
−0.76	22.36	27.64	22.36	.2989
−0.75	22.66	27.34	22.66	.3011
−0.74	22.96	27.04	22.96	.3034
−0.73	23.27	26.73	23.27	.3056
−0.72	23.58	26.42	23.58	.3079
−0.71	23.89	26.11	23.89	.3101
−0.70	24.20	25.80	24.20	.3123
−0.69	24.51	25.49	24.51	.3144
−0.68	24.83	25.17	24.83	.3166
−0.67	25.14	24.86	25.14	.3187
−0.66	25.46	24.54	25.46	.3209
−0.65	25.78	24.22	25.78	.3230
−0.64	26.11	23.89	26.11	.3251
−0.63	26.43	23.56	26.43	.3271
−0.62	26.76	23.24	26.76	.3292
−0.61	27.09	22.91	27.09	.3312
−0.60	27.43	22.57	27.43	.3332
−0.59	27.76	22.24	27.76	.3352
−0.58	28.10	21.90	28.10	.3372
−0.57	28.43	21.57	28.43	.3391
−0.56	28.77	21.23	28.77	.3410
−0.55	29.12	20.88	29.12	.3429
−0.54	29.46	20.54	29.46	.3448
−0.53	29.81	20.19	29.81	.3467
−0.52	30.15	19.85	30.15	.3485
−0.51	30.50	19.50	30.50	.3503
−0.50	30.85	19.15	30.85	.3521
−0.49	31.21	18.79	31.21	.3538
−0.48	31.56	18.44	31.56	.3555
−0.47	31.92	18.08	31.92	.3572
−0.46	32.28	17.72	32.28	.3589
−0.45	32.64	17.36	32.64	.3605
−0.44	33.00	17.00	33.00	.3621
−0.43	33.36	16.64	33.36	.3637
−0.42	33.72	16.28	33.72	.3653
−0.41	34.09	15.91	34.09	.3668
−0.40	34.46	15.54	34.46	.3683
−0.39	34.83	15.17	34.83	.3697
−0.38	35.20	14.80	35.20	.3712
−0.37	35.57	14.43	35.57	.3725
−0.36	35.94	14.06	35.94	.3739
−0.35	36.32	13.68	36.32	.3752
−0.34	36.69	13.31	36.69	.3765
−0.33	37.07	12.93	37.07	.3778
−0.32	37.45	12.55	37.45	.3790
−0.31	37.83	12.17	37.83	.3802

Table A   (*Continued*)

z	(A) Percentile rank (percent of area below z)	(B) Percent of area from mean to z	(C) Percent of area beyond z	(D) Height of ordinate at z
−0.30	38.21	11.79	38.21	.3814
−0.29	38.59	11.41	38.59	.3825
−0.28	38.97	11.03	38.97	.3836
−0.27	39.36	10.64	39.36	.3847
−0.26	39.74	10.26	39.74	.3857
−0.25	40.13	09.87	40.13	.3867
−0.24	40.52	09.48	40.52	.3876
−0.23	40.90	09.10	40.90	.3885
−0.22	41.29	08.71	41.29	.3894
−0.21	41.68	08.32	41.68	.3902
−0.20	42.07	07.93	42.07	.3910
−0.19	42.47	07.53	42.47	.3918
−0.18	44.86	07.14	42.86	.3925
−0.17	43.25	06.75	43.25	.3932
−0.16	43.64	06.36	43.64	.3939
−0.15	44.04	05.96	44.04	.3945
−0.14	44.43	05.57	44.43	.3951
−0.13	44.83	05.17	44.83	.3956
−0.12	45.22	04.78	45.22	.3961
−0.11	45.62	04.38	45.62	.3965
−0.10	46.02	03.98	46.02	.3970
−0.09	46.41	03.59	46.41	.3973
−0.08	46.81	03.19	46.81	.3977
−0.07	47.21	02.79	47.21	.3980
−0.06	47.61	02.39	47.61	.3982
−0.05	48.01	01.99	48.01	.3984
−0.04	48.40	01.60	48.40	.3986
−0.03	48.80	01.20	48.80	.3988
−0.02	49.20	00.80	49.20	.3989
−0.01	49.60	00.40	49.60	.3989
0.00	50.00	00.00	50.00	.3989
0.01	50.40	00.40	49.60	.3989
0.02	50.80	00.80	49.20	.3989
0.03	51.20	01.20	48.80	.3988
0.04	51.60	01.60	48.40	.3986
0.05	51.99	01.99	48.01	.3984
0.06	52.39	02.39	47.61	.3982
0.07	52.79	02.79	47.21	.3980
0.08	53.19	03.19	46.81	.3977
0.09	53.59	03.59	46.41	.3973
0.10	53.98	03.98	46.02	.3970
0.11	54.38	04.38	45.62	.3965
0.12	54.78	04.78	45.22	.3961
0.13	55.17	05.17	44.83	.3956
0.14	55.57	05.57	44.43	.3951
0.15	55.96	05.96	44.04	.3945
0.16	56.36	06.36	43.64	.3939
0.17	56.75	06.75	43.25	.3932
0.18	57.14	07.14	42.86	.3925
0.19	57.53	07.53	42.47	.3918

Table A   (*Continued*)

z	(A) Percentile rank (percent of area below z)	(B) Percent of area from mean to z	(C) Percent of area beyond z	(D) Height of ordinate at z
0.20	57.93	07.93	42.07	.3910
0.21	58.32	08.32	41.68	.3902
0.22	58.71	08.71	41.29	.3894
0.23	59.10	09.10	40.90	.3885
0.24	59.48	09.48	40.52	.3876
0.25	59.87	09.87	40.13	.3867
0.26	60.26	10.26	39.74	.3857
0.27	60.64	10.64	39.36	.3847
0.28	61.03	11.03	38.97	.3836
0.29	61.41	11.41	38.59	.3825
0.30	61.79	11.79	38.21	.3814
0.31	62.17	12.17	37.83	.3802
0.32	62.55	12.55	37.45	.3790
0.33	62.93	12.93	37.07	.3778
0.34	63.31	13.31	36.69	.3765
0.35	63.68	13.68	36.32	.3752
0.36	64.06	14.06	35.94	.3739
0.37	64.43	14.43	35.57	.3725
0.38	64.80	14.80	35.20	.3712
0.39	65.17	15.17	34.83	.3697
0.40	65.54	15.54	34.46	.3683
0.41	65.91	15.91	34.09	.3668
0.42	66.28	16.28	33.72	.3653
0.43	66.64	16.64	33.36	.3637
0.44	67.00	17.00	33.00	.3621
0.45	67.36	17.36	32.64	.3605
0.46	67.72	17.72	32.28	.3589
0.47	68.08	18.08	31.92	.3572
0.48	68.44	18.44	31.56	.3555
0.49	68.79	18.79	31.21	.3538
0.50	69.15	19.15	30.85	.3521
0.51	69.50	19.50	30.50	.3503
0.52	69.85	19.85	30.15	.3485
0.53	70.19	20.19	29.81	.3467
0.54	70.54	20.54	29.46	.3448
0.55	70.88	20.88	29.12	.3429
0.56	71.23	21.23	28.77	.3410
0.57	71.57	21.57	28.43	.3391
0.58	71.90	21.90	28.10	.3372
0.59	72.24	22.24	27.76	.3352
0.60	72.57	22.57	27.43	.3332
0.61	72.91	22.91	27.09	.3312
0.62	73.24	23.24	26.76	.3292
0.63	73.57	23.57	26.43	.3271
0.64	73.89	23.89	26.11	.3251
0.65	74.22	24.22	25.78	.3230
0.66	74.54	24.54	25.46	.3209
0.67	74.86	24.86	25.14	.3187
0.68	75.17	25.17	24.83	.3166
0.69	75.49	25.49	24.51	.3144

Table A   *(Continued)*

z	(A) Percentile rank (percent of area below z)	(B) Percent of area from mean to z	(C) Percent of area beyond z	(D) Height of ordinate at z
0.70	75.80	25.80	24.20	.3123
0.71	76.11	26.11	23.89	.3101
0.72	76.42	26.42	23.58	.3079
0.73	76.73	26.73	23.27	.3056
0.74	77.04	27.04	22.96	.3034
0.75	77.34	27.34	22.66	.3011
0.76	77.64	27.64	22.36	.2989
0.77	77.94	27.94	22.06	.2966
0.78	78.23	28.23	21.77	.2943
0.79	78.52	28.52	21.48	.2920
0.80	78.81	28.81	21.19	.2897
0.81	79.10	29.10	20.90	.2874
0.82	79.39	29.39	20.61	.2850
0.83	79.67	29.67	20.33	.2827
0.84	79.95	29.95	20.05	.2803
0.85	80.23	30.23	19.77	.2780
0.86	80.51	30.51	19.49	.2756
0.87	80.78	30.78	19.22	.2732
0.88	81.06	31.06	18.94	.2709
0.89	81.33	31.33	18.67	.2685
0.90	81.59	31.59	18.41	.2661
0.91	81.86	31.86	18.14	.2637
0.92	82.12	32.12	17.88	.2613
0.93	82.38	32.38	17.62	.2589
0.94	82.64	32.64	17.36	.2565
0.95	82.89	32.89	17.11	.2541
0.96	83.15	33.15	16.85	.2516
0.97	83.40	33.40	16.60	.2492
0.98	83.65	33.65	16.35	.2468
0.99	83.89	33.89	16.11	.2444
1.00	84.13	34.13	15.87	.2420
1.01	84.38	34.38	15.62	.2396
1.02	84.61	34.61	15.39	.2371
1.03	84.85	34.85	15.15	.2347
1.04	85.08	35.08	14.92	.2323
1.05	85.31	35.31	14.69	.2299
1.06	85.54	35.54	14.46	.2275
1.07	85.77	35.77	14.23	.2251
1.08	85.99	35.99	14.01	.2227
1.09	86.21	36.21	13.79	.2203
1.10	86.43	36.43	13.57	.2179
1.11	86.65	36.65	13.35	.2155
1.12	86.86	36.86	13.14	.2131
1.13	87.08	37.08	12.92	.2107
1.14	87.29	37.29	12.71	.2083
1.15	87.49	37.49	12.51	.2059
1.16	87.70	37.70	12.30	.2036
1.17	87.90	37.90	12.10	.2012
1.18	88.10	38.10	11.90	.1989
1.19	88.30	38.30	11.70	.1965

Table A   (*Continued*)

z	(A) Percentile rank (percent of area below z)	(B) Percent of area from mean to z	(C) Percent of area beyond z	(D) Height of ordinate at z
1.20	88.49	38.49	11.51	.1942
1.21	88.69	38.69	11.31	.1919
1.22	88.88	38.88	11.12	.1895
1.23	89.07	39.07	10.93	.1872
1.24	89.25	39.25	10.75	.1849
1.25	89.44	39.44	10.56	.1826
1.26	89.62	39.62	10.38	.1804
1.27	89.80	39.80	10.20	.1781
1.28	89.97	39.97	10.03	.1758
1.29	90.15	40.15	09.85	.1736
1.30	90.32	40.32	09.68	.1714
1.31	90.49	40.49	09.51	.1691
1.32	90.66	40.66	09.34	.1669
1.33	90.82	40.82	09.18	.1647
1.34	90.99	40.99	09.01	.1626
1.35	91.15	41.15	08.85	.1604
1.36	91.31	41.31	08.69	.1582
1.37	91.47	41.47	08.53	.1561
1.38	91.62	41.62	08.38	.1539
1.39	91.77	41.77	08.23	.1518
1.40	91.92	41.92	08.08	.1497
1.41	92.07	42.07	07.93	.1476
1.42	92.22	42.22	07.78	.1456
1.43	92.36	42.36	07.64	.1435
1.44	92.51	42.51	07.49	.1415
1.45	92.65	42.65	07.35	.1394
1.46	92.79	42.79	07.21	.1374
1.47	92.92	42.92	07.08	.1354
1.48	93.06	43.06	06.94	.1334
1.49	93.19	43.19	06.81	.1315
1.50	93.32	43.32	06.68	.1295
1.51	93.45	43.45	06.55	.1276
1.52	93.57	43.57	06.43	.1257
1.53	93.70	43.70	06.30	.1238
1.54	93.82	43.82	06.18	.1219
1.55	93.94	43.94	06.06	.1200
1.56	94.06	44.06	05.94	.1182
1.57	94.18	44.18	05.82	.1163
1.58	94.29	44.29	05.71	.1145
1.59	94.41	44.41	05.59	.1127
1.60	94.52	44.52	05.48	.1109
1.61	94.63	44.63	05.37	.1092
1.62	94.74	44.74	05.26	.1074
1.63	94.84	44.84	05.16	.1057
1.64	94.95	44.95	05.05	.1040
1.65	95.05	45.05	04.95	.1023
1.66	95.15	45.15	04.85	.1006
1.67	95.25	45.25	04.75	.0989
1.68	95.35	45.35	04.65	.0973
1.69	95.45	45.45	04.55	.0957

Table A *(Continued)*

z	(A) Percentile rank (percent of area below z)	(B) Percent of area from mean to z	(C) Percent of area beyond z	(D) Height of ordinate at z
1.70	95.54	45.54	04.46	.0940
1.71	95.64	45.64	04.36	.0925
1.72	95.73	45.73	04.27	.0909
1.73	95.82	45.82	04.18	.0893
1.74	95.91	45.91	04.09	.0878
1.75	95.99	45.99	04.01	.0863
1.76	96.08	46.08	03.92	.0848
1.77	96.16	46.16	03.84	.0833
1.78	96.25	46.25	03.75	.0818
1.79	96.33	46.33	03.67	.0804
1.80	96.41	46.41	03.59	.0790
1.81	96.49	46.49	03.51	.0775
1.82	96.56	46.56	03.44	.0761
1.83	96.64	46.64	03.36	.0748
1.84	96.71	46.71	03.29	.0734
1.85	96.78	46.78	03.22	.0721
1.86	96.86	46.86	03.14	.0707
1.87	96.93	46.93	03.07	.0694
1.88	96.99	46.99	03.01	.0681
1.89	97.06	47.06	02.94	.0669
1.90	97.13	47.13	02.87	.0656
1.91	97.19	47.19	02.81	.0644
1.92	97.26	47.26	02.74	.0632
1.93	97.32	47.32	02.68	.0620
1.94	97.38	47.38	02.62	.0608
1.95	97.44	47.44	02.56	.0596
1.96	97.50	47.50	02.50	.0584
1.97	97.56	47.56	02.44	.0573
1.98	97.61	47.61	02.39	.0562
1.99	97.67	47.67	02.33	.0551
2.00	97.72	47.72	02.28	.0540
2.01	97.78	47.78	02.22	.0529
2.02	97.83	47.83	02.17	.0519
2.03	97.88	47.88	02.12	.0508
2.04	97.93	47.93	02.07	.0498
2.05	97.98	47.98	02.02	.0488
2.06	98.03	48.03	01.97	.0478
2.07	98.08	48.08	01.92	.0468
2.08	98.12	48.12	01.88	.0459
2.09	98.17	48.17	01.83	.0449
2.10	98.21	48.21	01.79	.0440
2.11	98.26	48.26	01.74	.0431
2.12	98.30	48.30	01.70	.0422
2.13	98.34	48.34	01.66	.0413
2.14	98.38	48.38	01.62	.0404
2.15	98.42	48.42	01.58	.0396
2.16	98.46	48.46	01.54	.0387
2.17	98.50	48.50	01.50	.0379
2.18	98.54	48.54	01.46	.0371
2.19	98.57	48.57	01.43	.0363

Table A *(Continued)*

z	(A) Percentile rank (percent of area below z)	(B) Percent of area from mean to z	(C) Percent of area beyond z	(D) Height of ordinate at z
2.20	98.61	48.61	01.39	.0355
2.21	98.64	48.64	01.36	.0347
2.22	98.68	48.68	01.32	.0339
2.23	98.71	48.71	01.29	.0332
2.24	98.75	48.75	01.25	.0325
2.25	98.78	48.78	01.22	.0317
2.26	98.81	48.81	01.19	.0310
2.27	98.84	48.84	01.16	.0303
2.28	98.87	48.87	01.13	.0297
2.29	98.90	48.90	01.10	.0290
2.30	98.93	48.93	01.07	.0283
2.31	98.96	48.96	01.04	.0277
2.32	98.98	48.98	01.02	.0270
2.33	99.01	49.01	00.99	.0264
2.34	99.04	49.04	00.96	.0258
2.35	99.06	49.06	00.94	.0252
2.36	99.09	49.09	00.91	.0246
2.37	99.11	49.11	00.89	.0241
2.38	99.13	49.13	00.87	.0235
2.39	99.16	49.16	00.84	.0229
2.40	99.18	49.18	00.82	.0224
2.41	99.20	49.20	00.80	.0219
2.42	99.22	49.22	00.78	.0213
2.43	99.25	49.25	00.75	.0208
2.44	99.27	49.27	00.73	.0203
2.45	99.29	49.29	00.71	.0198
2.46	99.31	49.31	00.69	.0194
2.47	99.32	49.32	00.68	.0189
2.48	99.34	49.34	00.66	.0184
2.49	99.36	49.36	00.64	.0180
2.50	99.38	49.38	00.62	.0175
2.51	99.40	49.40	00.60	.0171
2.52	99.41	49.41	00.59	.0167
2.53	99.43	49.43	00.57	.0163
2.54	99.45	49.45	00.55	.0158
2.55	99.46	49.46	00.54	.0154
2.56	99.48	49.48	00.52	.0151
2.57	99.49	49.49	00.51	.0147
2.58	99.51	49.51	00.49	.0143
2.59	99.52	49.52	00.48	.0139
2.60	99.53	49.53	00.47	.0136
2.61	99.55	49.55	00.45	.0132
2.62	99.56	49.56	00.44	.0129
2.63	99.57	49.57	00.43	.0126
2.64	99.59	49.59	00.41	.0122
2.65	99.60	49.60	00.40	.0119
2.66	99.61	49.61	00.39	.0116
2.67	99.62	49.62	00.38	.0113
2.68	99.63	49.63	00.37	.0110
2.69	99.64	49.64	00.36	.0107

Table A *(Continued)*

z	(A) Percentile rank (percent of area below z)	(B) Percent of area from mean to z	(C) Percent of area beyond z	(D) Height of ordinate at z
2.70	99.65	49.65	00.35	.0104
2.71	99.66	49.66	00.34	.0101
2.72	99.67	49.67	00.33	.0099
2.73	99.68	49.68	00.32	.0096
2.74	99.69	49.69	00.31	.0093
2.75	99.70	49.70	00.30	.0091
2.76	99.71	49.71	00.29	.0088
2.77	99.72	49.72	00.28	.0086
2.78	99.73	49.73	00.27	.0084
2.79	99.74	49.74	00.26	.0081
2.80	99.74	49.74	00.26	.0079
2.81	99.75	49.75	00.25	.0077
2.82	99.76	49.76	00.24	.0075
2.83	99.77	49.77	00.23	.0073
2.84	99.77	49.77	00.23	.0071
2.85	99.78	49.78	00.22	.0069
2.86	99.79	49.79	00.21	.0067
2.87	99.79	49.79	00.21	.0065
2.88	99.80	49.80	00.20	.0063
2.89	99.81	49.81	00.19	.0061
2.90	99.81	49.81	00.19	.0060
2.91	99.82	49.82	00.18	.0058
2.92	99.82	49.82	00.18	.0056
2.93	99.83	49.83	00.17	.0055
2.94	99.84	49.84	00.16	.0053
2.95	99.84	49.84	00.16	.0051
2.96	99.85	49.85	00.15	.0051
2.97	99.85	49.85	00.15	.0048
2.98	99.86	49.86	00.14	.0047
2.99	99.86	49.86	00.14	.0046
3.00	99.87	49.87	00.13	.0044
3.01	99.87	49.87	00.13	.0043
3.02	99.87	49.87	00.13	.0042
3.03	99.88	49.88	00.12	.0040
3.04	99.88	49.88	00.12	.0039
3.05	99.89	49.89	00.11	.0038
3.06	99.89	49.89	00.11	.0037
3.07	99.89	49.89	00.11	.0036
3.08	99.90	49.90	00.10	.0035
3.09	99.90	49.90	00.10	.0034
3.10	99.90	49.90	00.10	.0033
3.11	99.91	49.91	00.09	.0032
3.12	99.91	49.91	00.09	.0031
3.13	99.91	49.91	00.09	.0030
3.14	99.92	49.92	00.08	.0029
3.15	99.92	49.92	00.08	.0028
3.16	99.92	49.92	00.08	.0027
3.17	99.92	49.92	00.08	.0026
3.18	99.93	49.93	00.07	.0025
3.19	99.93	49.93	00.07	.0025

Table A   *(Continued)*

z	(A) Percentile rank (percent of area below z)	(B) Percent of area from mean to z	(C) Percent of area beyond z	(D) Height of ordinate at z
3.20	99.93	49.93	00.07	.0024
3.21	99.93	49.93	00.07	.0023
3.22	99.94	49.94	00.06	.0022
3.23	99.94	49.94	00.06	.0022
3.24	99.94	49.94	00.06	.0021
3.30	99.95	49.95	00.05	.0017
3.40	99.97	49.97	00.03	.0012
3.50	99.98	49.98	00.02	.0009
3.60	99.98	49.98	00.02	.0006
3.70	99.99	49.99	00.01	.0004

Table B Critical values of t

For any given df, the table shows the values of t corresponding to various levels of probability. Obtained t is significant at a given level if it is equal to or <u>greater than</u> the value shown in the table.

df	Level of significance for one-tailed test					
	.10	.05	.025	.01	.005	.0005
	Level of significance for two-tailed test					
	.20	.10	.05	.02	.01	.001
1	3.078	6.314	12.706	31.821	63.657	636.619
2	1.886	2.920	4.303	6.965	9.925	31.598
3	1.638	2.353	3.182	4.541	5.841	12.941
4	1.533	2.132	2.776	3.747	4.604	8.610
5	1.476	2.015	2.571	3.365	4.032	6.859
6	1.440	1.943	2.447	3.143	3.707	5.959
7	1.415	1.895	2.365	2.998	3.499	5.405
8	1.397	1.860	2.306	2.896	3.355	5.041
9	1.383	1.833	2.262	2.821	3.250	4.781
10	1.372	1.812	2.228	2.764	3.169	4.587
11	1.363	1.796	2.201	2.718	3.106	4.437
12	1.356	1.782	2.179	2.681	3.055	4.318
13	1.350	1.771	2.160	2.650	3.012	4.221
14	1.345	1.761	2.145	2.624	2.977	4.140
15	1.341	1.753	2.131	2.602	2.947	4.073
16	1.337	1.746	2.120	2.583	2.921	4.015
17	1.333	1.740	2.110	2.567	2.898	3.965
18	1.330	1.734	2.101	2.552	2.878	3.922
19	1.328	1.729	2.093	2.539	2.861	3.883
20	1.325	1.725	2.086	2.528	2.845	3.850
21	1.323	1.721	2.080	2.518	2.831	3.819
22	1.321	1.717	2.074	2.508	2.819	3.792
23	1.319	1.714	2.069	2.500	2.807	3.767
24	1.318	1.711	2.064	2.492	2.797	3.745
25	1.316	1.708	2.060	2.485	2.787	3.725
26	1.315	1.706	2.056	2.479	2.779	3.707
27	1.314	1.703	2.052	2.473	2.771	3.690
28	1.313	1.701	2.048	2.467	2.763	3.674
29	1.311	1.699	2.045	2.462	2.756	3.659
30	1.310	1.697	2.042	2.457	2.750	3.646
40	1.303	1.684	2.021	2.423	2.704	3.551
60	1.296	1.671	2.000	2.390	2.660	3.460
120	1.289	1.658	1.980	2.358	2.617	3.373
∞	1.282	1.645	1.960	2.326	2.576	3.291

Table C Critical values of F

The obtained F is significant at a given level if it is equal to or greater than the value shown in the table.
0.05 (light row) and 0.01 (dark row) points for the distribution of F

Degrees of freedom for numerator (each cell: 0.05 value / 0.01 value)

df denom	1	2	3	4	5	6	7	8	9	10	11	12	14	16	20	24	30	40	50	75	100	200	500	∞
1	161 / 4052	200 / 4999	216 / 5403	225 / 5625	230 / 5764	234 / 5859	237 / 5928	239 / 5981	241 / 6022	242 / 6056	243 / 6082	244 / 6106	245 / 6142	246 / 6169	248 / 6208	249 / 6234	250 / 6258	251 / 6286	252 / 6302	253 / 6323	253 / 6334	254 / 6352	254 / 6361	254 / 6366
2	18.51 / 98.49	19.00 / 99.01	19.16 / 99.17	19.25 / 99.25	19.30 / 99.30	19.33 / 99.33	19.36 / 99.34	19.37 / 99.36	19.38 / 99.38	19.39 / 99.40	19.40 / 99.41	19.41 / 99.42	19.42 / 99.43	19.43 / 99.44	19.44 / 99.45	19.45 / 99.46	19.46 / 99.47	19.47 / 99.48	19.47 / 99.48	19.48 / 99.49	19.49 / 99.49	19.49 / 99.49	19.50 / 99.50	19.50 / 99.50
3	10.13 / 34.12	9.55 / 30.81	9.28 / 29.46	9.12 / 28.71	9.01 / 28.24	8.94 / 27.91	8.88 / 27.67	8.84 / 27.49	8.81 / 27.34	8.78 / 27.23	8.76 / 27.13	8.74 / 27.05	8.71 / 26.92	8.69 / 26.83	8.66 / 26.69	8.64 / 26.60	8.62 / 26.50	8.60 / 26.41	8.58 / 26.30	8.57 / 26.27	8.56 / 26.23	8.54 / 26.18	8.54 / 26.14	8.53 / 26.12
4	7.71 / 21.20	6.94 / 18.00	6.59 / 16.69	6.39 / 15.98	6.26 / 15.52	6.16 / 15.21	6.09 / 14.98	6.04 / 14.80	6.00 / 14.66	5.96 / 14.54	5.93 / 14.45	5.91 / 14.37	5.87 / 14.24	5.84 / 14.15	5.80 / 14.02	5.77 / 13.93	5.74 / 13.83	5.71 / 13.74	5.70 / 13.69	5.68 / 13.61	5.66 / 13.57	5.65 / 13.52	5.64 / 13.48	5.63 / 13.46
5	6.61 / 16.26	5.79 / 13.27	5.41 / 12.06	5.19 / 11.39	5.05 / 10.97	4.95 / 10.67	4.88 / 10.45	4.82 / 10.27	4.78 / 10.15	4.74 / 10.05	4.70 / 9.96	4.68 / 9.89	4.64 / 9.77	4.60 / 9.68	4.56 / 9.55	4.53 / 9.47	4.50 / 9.38	4.46 / 9.29	4.44 / 9.24	4.42 / 9.17	4.40 / 9.13	4.38 / 9.07	4.37 / 9.04	4.36 / 9.02
6	5.99 / 13.74	5.14 / 10.92	4.76 / 9.78	4.53 / 9.15	4.39 / 8.75	4.28 / 8.47	4.21 / 8.26	4.15 / 8.10	4.10 / 7.98	4.06 / 7.87	4.03 / 7.79	4.00 / 7.72	3.96 / 7.60	3.92 / 7.52	3.87 / 7.39	3.84 / 7.31	3.81 / 7.23	3.77 / 7.14	3.75 / 7.09	3.72 / 7.02	3.71 / 6.99	3.69 / 6.94	3.68 / 6.90	3.67 / 6.88
7	5.59 / 12.25	4.74 / 9.55	4.35 / 8.45	4.12 / 7.85	3.97 / 7.46	3.87 / 7.19	3.79 / 7.00	3.73 / 6.84	3.68 / 6.71	3.63 / 6.62	3.60 / 6.54	3.57 / 6.47	3.52 / 6.35	3.49 / 6.27	3.44 / 6.15	3.41 / 6.07	3.38 / 5.98	3.34 / 5.90	3.32 / 5.85	3.29 / 5.78	3.28 / 5.75	3.25 / 5.70	3.24 / 5.67	3.23 / 5.65
8	5.32 / 11.26	4.46 / 8.65	4.07 / 7.59	3.84 / 7.01	3.69 / 6.63	3.58 / 6.37	3.50 / 6.19	3.44 / 6.03	3.39 / 5.91	3.34 / 5.82	3.31 / 5.74	3.28 / 5.67	3.23 / 5.56	3.20 / 5.48	3.15 / 5.36	3.12 / 5.28	3.08 / 5.20	3.05 / 5.11	3.03 / 5.06	3.00 / 5.00	2.98 / 4.96	2.96 / 4.91	2.94 / 4.88	2.93 / 4.86
9	5.12 / 10.56	4.26 / 8.02	3.86 / 6.99	3.63 / 6.42	3.48 / 6.06	3.37 / 5.80	3.29 / 5.62	3.23 / 5.47	3.18 / 5.35	3.13 / 5.26	3.10 / 5.18	3.07 / 5.11	3.02 / 5.00	2.98 / 4.92	2.93 / 4.80	2.90 / 4.73	2.86 / 4.64	2.82 / 4.56	2.80 / 4.51	2.77 / 4.45	2.76 / 4.41	2.73 / 4.36	2.72 / 4.33	2.71 / 4.31
10	4.96 / 10.04	4.10 / 7.56	3.71 / 6.55	3.48 / 5.99	3.33 / 5.64	3.22 / 5.39	3.14 / 5.21	3.07 / 5.06	3.02 / 4.95	2.97 / 4.85	2.94 / 4.78	2.91 / 4.71	2.86 / 4.60	2.82 / 4.52	2.77 / 4.41	2.74 / 4.33	2.70 / 4.25	2.67 / 4.17	2.64 / 4.12	2.61 / 4.05	2.59 / 4.01	2.56 / 3.96	2.55 / 3.93	2.54 / 3.91
11	4.84 / 9.65	3.98 / 7.20	3.59 / 6.22	3.36 / 5.67	3.20 / 5.32	3.09 / 5.07	3.01 / 4.88	2.95 / 4.74	2.90 / 4.63	2.86 / 4.54	2.82 / 4.46	2.79 / 4.40	2.74 / 4.29	2.70 / 4.21	2.65 / 4.10	2.61 / 4.02	2.57 / 3.94	2.53 / 3.86	2.50 / 3.80	2.47 / 3.74	2.45 / 3.70	2.42 / 3.66	2.41 / 3.62	2.40 / 3.60
12	4.75 / 9.33	3.88 / 6.93	3.49 / 5.95	3.26 / 5.41	3.11 / 5.06	3.00 / 4.82	2.92 / 4.65	2.85 / 4.50	2.80 / 4.39	2.76 / 4.30	2.72 / 4.22	2.69 / 4.16	2.64 / 4.05	2.60 / 3.98	2.54 / 3.86	2.50 / 3.78	2.46 / 3.70	2.42 / 3.61	2.40 / 3.56	2.36 / 3.49	2.35 / 3.46	2.32 / 3.41	2.31 / 3.38	2.30 / 3.36
13	4.67 / 9.07	3.80 / 6.70	3.41 / 5.74	3.18 / 5.20	3.02 / 4.86	2.92 / 4.62	2.84 / 4.44	2.77 / 4.30	2.72 / 4.19	2.67 / 4.10	2.63 / 4.02	2.60 / 3.96	2.55 / 3.85	2.51 / 3.78	2.46 / 3.67	2.42 / 3.59	2.38 / 3.51	2.34 / 3.42	2.32 / 3.37	2.28 / 3.30	2.26 / 3.27	2.24 / 3.21	2.22 / 3.18	2.21 / 3.16
14	4.60 / 8.86	3.74 / 6.51	3.34 / 5.56	3.11 / 5.03	2.96 / 4.69	2.85 / 4.46	2.77 / 4.28	2.70 / 4.14	2.65 / 4.03	2.60 / 3.94	2.56 / 3.86	2.53 / 3.80	2.48 / 3.70	2.44 / 3.62	2.39 / 3.51	2.35 / 3.43	2.31 / 3.34	2.27 / 3.26	2.24 / 3.21	2.21 / 3.14	2.19 / 3.11	2.16 / 3.06	2.14 / 3.02	2.13 / 3.00
15	4.54 / 8.68	3.68 / 6.36	3.29 / 5.42	3.06 / 4.89	2.90 / 4.56	2.79 / 4.32	2.70 / 4.14	2.64 / 4.00	2.59 / 3.89	2.55 / 3.80	2.51 / 3.73	2.48 / 3.67	2.43 / 3.56	2.39 / 3.48	2.33 / 3.36	2.29 / 3.29	2.25 / 3.20	2.21 / 3.12	2.18 / 3.07	2.15 / 3.00	2.12 / 2.97	2.10 / 2.92	2.08 / 2.89	2.07 / 2.87

Degrees of freedom for denominator

Table C.   Critical values of F   139

Each cell lists the 0.05 critical value (upper) and the 0.01 critical value (lower). Column headers (degrees of freedom for numerator) are not shown on this portion of the page.

df (denom.)																								
16	4.49/8.53	3.63/6.23	3.24/5.29	3.01/4.77	2.85/4.44	2.74/4.20	2.66/4.03	2.59/3.89	2.54/3.78	2.49/3.69	2.45/3.61	2.42/3.55	2.37/3.45	2.33/3.37	2.28/3.25	2.24/3.18	2.20/3.10	2.16/3.01	2.13/2.96	2.09/2.89	2.07/2.86	2.04/2.80	2.02/2.77	2.01/2.75
17	4.45/8.40	3.59/6.11	3.20/5.18	2.96/4.67	2.81/4.34	2.70/4.10	2.62/3.93	2.55/3.79	2.50/3.68	2.45/3.59	2.41/3.52	2.38/3.45	2.33/3.35	2.29/3.27	2.23/3.16	2.19/3.08	2.15/3.00	2.11/2.92	2.08/2.86	2.04/2.79	2.02/2.76	1.99/2.70	1.97/2.67	1.96/2.65
18	4.41/8.28	3.55/6.01	3.16/5.09	2.93/4.58	2.77/4.25	2.66/4.01	2.58/3.85	2.51/3.71	2.46/3.60	2.41/3.51	2.37/3.44	2.34/3.37	2.29/3.27	2.25/3.19	2.19/3.07	2.15/3.00	2.11/2.91	2.07/2.83	2.04/2.78	2.00/2.71	1.98/2.68	1.95/2.62	1.93/2.59	1.92/2.57
19	4.38/8.18	3.52/5.93	3.13/5.01	2.90/4.50	2.74/4.17	2.63/3.94	2.55/3.77	2.48/3.63	2.43/3.52	2.38/3.43	2.34/3.36	2.31/3.30	2.26/3.19	2.21/3.12	2.15/3.00	2.11/2.92	2.07/2.84	2.02/2.76	2.00/2.70	1.96/2.63	1.94/2.60	1.91/2.54	1.90/2.51	1.88/2.49
20	4.35/8.10	3.49/5.85	3.10/4.94	2.87/4.43	2.71/4.10	2.60/3.87	2.52/3.71	2.45/3.56	2.40/3.45	2.35/3.37	2.31/3.30	2.28/3.23	2.23/3.13	2.18/3.05	2.12/2.94	2.08/2.86	2.04/2.77	1.99/2.69	1.96/2.63	1.92/2.56	1.90/2.53	1.87/2.47	1.85/2.44	1.84/2.42
21	4.32/8.02	3.47/5.78	3.07/4.87	2.84/4.37	2.68/4.04	2.57/3.81	2.49/3.65	2.42/3.51	2.37/3.40	2.32/3.31	2.28/3.24	2.25/3.17	2.20/3.07	2.15/2.99	2.09/2.88	2.05/2.80	2.00/2.72	1.96/2.63	1.93/2.58	1.89/2.51	1.87/2.47	1.84/2.42	1.82/2.38	1.81/2.36
22	4.30/7.94	3.44/5.72	3.05/4.82	2.82/4.31	2.66/3.99	2.55/3.76	2.47/3.59	2.40/3.45	2.35/3.35	2.30/3.26	2.26/3.18	2.23/3.12	2.18/3.02	2.13/2.94	2.07/2.83	2.03/2.75	1.98/2.67	1.93/2.58	1.91/2.53	1.87/2.46	1.84/2.42	1.81/2.37	1.80/2.33	1.78/2.31
23	4.28/7.88	3.42/5.66	3.03/4.76	2.80/4.26	2.64/3.94	2.53/3.71	2.45/3.54	2.38/3.41	2.32/3.30	2.28/3.21	2.24/3.14	2.20/3.07	2.14/2.97	2.10/2.89	2.04/2.78	2.00/2.70	1.96/2.62	1.91/2.53	1.88/2.48	1.84/2.41	1.82/2.37	1.79/2.32	1.77/2.28	1.76/2.26
24	4.26/7.82	3.40/5.61	3.01/4.72	2.78/4.22	2.62/3.90	2.51/3.67	2.43/3.50	2.36/3.36	2.30/3.25	2.26/3.17	2.22/3.09	2.18/3.03	2.13/2.93	2.09/2.85	2.02/2.74	1.98/2.66	1.94/2.58	1.89/2.49	1.86/2.44	1.82/2.36	1.80/2.33	1.76/2.27	1.74/2.23	1.73/2.21
25	4.24/7.77	3.38/5.57	2.99/4.68	2.76/4.18	2.60/3.86	2.49/3.63	2.41/3.46	2.34/3.32	2.28/3.21	2.24/3.13	2.20/3.05	2.16/2.99	2.11/2.89	2.06/2.81	2.00/2.70	1.96/2.62	1.92/2.54	1.87/2.45	1.84/2.40	1.80/2.32	1.77/2.29	1.74/2.23	1.72/2.19	1.71/2.17
26	4.22/7.72	3.37/5.53	2.96/4.64	2.74/4.14	2.59/3.82	2.47/3.59	2.39/3.42	2.32/3.29	2.27/3.17	2.22/3.09	2.18/3.02	2.15/2.96	2.10/2.86	2.05/2.77	1.99/2.66	1.95/2.58	1.90/2.50	1.85/2.41	1.82/2.36	1.78/2.28	1.76/2.25	1.72/2.19	1.70/2.15	1.69/2.13
27	4.21/7.68	3.35/5.49	2.96/4.60	2.73/4.11	2.57/3.79	2.46/3.56	2.37/3.39	2.30/3.26	2.25/3.14	2.20/3.06	2.16/2.98	2.13/2.93	2.08/2.83	2.03/2.74	1.97/2.63	1.93/2.55	1.88/2.47	1.84/2.38	1.80/2.33	1.76/2.25	1.74/2.21	1.71/2.16	1.68/2.12	1.67/2.10
28	4.20/7.64	3.34/5.45	2.95/4.57	2.71/4.07	2.56/3.76	2.44/3.53	2.36/3.36	2.29/3.23	2.24/3.11	2.19/3.03	2.15/2.95	2.12/2.90	2.06/2.80	2.02/2.71	1.96/2.60	1.91/2.52	1.87/2.44	1.81/2.35	1.78/2.30	1.75/2.22	1.72/2.18	1.69/2.13	1.67/2.09	1.65/2.06
29	4.18/7.60	3.33/5.52	2.93/4.54	2.70/4.04	2.54/3.73	2.43/3.50	2.35/3.33	2.28/3.20	2.22/3.08	2.18/3.00	2.14/2.92	2.10/2.87	2.05/2.77	2.00/2.68	1.94/2.57	1.90/2.49	1.85/2.41	1.80/2.32	1.77/2.27	1.73/2.19	1.71/2.15	1.68/2.10	1.65/2.06	1.64/2.03
30	4.17/7.56	3.32/5.39	2.92/4.51	2.69/4.02	2.53/3.70	2.42/3.47	2.34/3.30	2.27/3.17	2.21/3.06	2.16/2.98	2.12/2.90	2.09/2.84	2.04/2.74	1.99/2.66	1.93/2.55	1.89/2.47	1.84/2.38	1.79/2.29	1.76/2.24	1.72/2.16	1.69/2.13	1.66/2.07	1.64/2.03	1.62/2.01

Degrees of freedom for denominator

Table C   (Continued)

0.05 (light row) and 0.01 (dark row) points for the distribution of F

Degrees of freedom for numerator

Denom. df	1	2	3	4	5	6	7	8	9	10	11	12	14	16	20	24	30	40	50	75	100	200	500	∞
32	4.15 / 7.50	3.30 / 5.34	2.90 / 4.46	2.67 / 3.97	2.51 / 3.66	2.40 / 3.42	2.32 / 3.25	2.25 / 3.12	2.19 / 3.01	2.14 / 2.94	2.10 / 2.86	2.07 / 2.80	2.02 / 2.70	1.97 / 2.62	1.91 / 2.51	1.86 / 2.42	1.82 / 2.34	1.76 / 2.25	1.74 / 2.20	1.69 / 2.12	1.67 / 2.08	1.64 / 2.02	1.61 / 1.98	1.59 / 1.96
34	4.13 / 7.44	3.28 / 5.29	2.88 / 4.42	2.65 / 3.93	2.49 / 3.61	2.38 / 3.38	2.30 / 3.21	2.23 / 3.08	2.17 / 2.97	2.12 / 2.89	2.08 / 2.82	2.05 / 2.76	2.00 / 2.66	1.95 / 2.58	1.89 / 2.47	1.84 / 2.38	1.80 / 2.30	1.74 / 2.21	1.71 / 2.15	1.67 / 2.08	1.64 / 2.04	1.61 / 1.98	1.59 / 1.94	1.57 / 1.91
36	4.11 / 7.39	3.26 / 5.25	2.86 / 4.38	2.63 / 3.89	2.48 / 3.58	2.36 / 3.35	2.28 / 3.18	2.21 / 3.04	2.15 / 2.94	2.10 / 2.86	2.06 / 2.78	2.03 / 2.72	1.98 / 2.62	1.93 / 2.54	1.87 / 2.43	1.82 / 2.35	1.78 / 2.26	1.72 / 2.17	1.69 / 2.12	1.65 / 2.04	1.62 / 2.00	1.59 / 1.94	1.56 / 1.90	1.55 / 1.87
38	4.10 / 7.35	3.25 / 5.21	2.85 / 4.34	2.62 / 3.86	2.46 / 3.54	2.35 / 3.32	2.26 / 3.15	2.19 / 3.02	2.14 / 2.91	2.09 / 2.82	2.05 / 2.75	2.02 / 2.69	1.96 / 2.59	1.92 / 2.51	1.85 / 2.40	1.80 / 2.32	1.76 / 2.22	1.71 / 2.14	1.67 / 2.08	1.63 / 2.00	1.60 / 1.97	1.57 / 1.90	1.54 / 1.86	1.53 / 1.84
40	4.08 / 7.31	3.23 / 5.18	2.84 / 4.31	2.61 / 3.83	2.45 / 3.51	2.34 / 3.29	2.25 / 3.12	2.18 / 2.99	2.12 / 2.88	2.07 / 2.80	2.04 / 2.73	2.00 / 2.66	1.95 / 2.56	1.90 / 2.49	1.84 / 2.37	1.79 / 2.29	1.74 / 2.20	1.69 / 2.11	1.66 / 2.05	1.61 / 1.97	1.59 / 1.94	1.55 / 1.88	1.53 / 1.84	1.51 / 1.81
42	4.07 / 7.27	3.22 / 5.15	2.83 / 4.29	2.59 / 3.80	2.44 / 3.49	2.32 / 3.26	2.24 / 3.10	2.17 / 2.96	2.11 / 2.86	2.06 / 2.77	2.02 / 2.70	1.99 / 2.64	1.94 / 2.54	1.89 / 2.46	1.82 / 2.35	1.78 / 2.26	1.73 / 2.17	1.68 / 2.08	1.64 / 2.02	1.60 / 1.94	1.57 / 1.91	1.54 / 1.85	1.51 / 1.80	1.49 / 1.78
44	4.06 / 7.24	3.21 / 5.12	2.82 / 4.26	2.58 / 3.78	2.43 / 3.46	2.31 / 3.24	2.23 / 3.07	2.16 / 2.94	2.10 / 2.84	2.05 / 2.75	2.01 / 2.68	1.98 / 2.62	1.92 / 2.52	1.88 / 2.44	1.81 / 2.32	1.76 / 2.24	1.72 / 2.15	1.66 / 2.06	1.63 / 2.00	1.58 / 1.92	1.56 / 1.88	1.52 / 1.82	1.50 / 1.78	1.48 / 1.75
46	4.05 / 7.21	3.20 / 5.10	2.81 / 4.24	2.57 / 3.76	2.42 / 3.44	2.30 / 3.22	2.22 / 3.05	2.14 / 2.92	2.09 / 2.82	2.04 / 2.73	2.00 / 2.66	1.97 / 2.60	1.91 / 2.50	1.87 / 2.42	1.80 / 2.30	1.75 / 2.22	1.71 / 2.13	1.65 / 2.04	1.62 / 1.98	1.57 / 1.90	1.54 / 1.86	1.51 / 1.80	1.48 / 1.76	1.46 / 1.72
48	4.04 / 7.19	3.19 / 5.08	2.80 / 4.22	2.56 / 3.74	2.41 / 3.42	2.30 / 3.20	2.21 / 3.04	2.14 / 2.90	2.08 / 2.80	2.03 / 2.71	1.99 / 2.64	1.96 / 2.58	1.90 / 2.48	1.86 / 2.40	1.79 / 2.28	1.74 / 2.20	1.70 / 2.11	1.64 / 2.02	1.61 / 1.96	1.56 / 1.88	1.53 / 1.84	1.50 / 1.78	1.47 / 1.73	1.45 / 1.70
50	4.03 / 7.17	3.18 / 5.06	2.79 / 4.20	2.56 / 3.72	2.40 / 3.41	2.29 / 3.18	2.20 / 3.02	2.13 / 2.88	2.07 / 2.78	2.02 / 2.70	1.98 / 2.62	1.95 / 2.56	1.90 / 2.46	1.85 / 2.39	1.78 / 2.26	1.74 / 2.18	1.69 / 2.10	1.63 / 2.00	1.60 / 1.94	1.55 / 1.86	1.52 / 1.82	1.48 / 1.76	1.46 / 1.71	1.44 / 1.68
55	4.02 / 7.12	3.17 / 5.01	2.78 / 4.16	2.54 / 3.68	2.38 / 3.37	2.27 / 3.15	2.18 / 2.98	2.11 / 2.85	2.05 / 2.75	2.00 / 2.66	1.97 / 2.59	1.93 / 2.53	1.88 / 2.43	1.83 / 2.35	1.76 / 2.23	1.72 / 2.15	1.67 / 2.06	1.61 / 1.96	1.58 / 1.90	1.52 / 1.82	1.50 / 1.78	1.46 / 1.71	1.43 / 1.66	1.41 / 1.64
60	4.00 / 7.08	3.15 / 4.98	2.76 / 4.13	2.52 / 3.65	2.37 / 3.34	2.25 / 3.12	2.17 / 2.95	2.10 / 2.82	2.04 / 2.72	1.99 / 2.63	1.95 / 2.56	1.92 / 2.50	1.86 / 2.40	1.81 / 2.32	1.75 / 2.20	1.70 / 2.12	1.65 / 2.03	1.59 / 1.93	1.56 / 1.87	1.50 / 1.79	1.48 / 1.74	1.44 / 1.68	1.41 / 1.63	1.39 / 1.60
65	3.99 / 7.04	3.14 / 4.95	2.75 / 4.10	2.51 / 3.62	2.36 / 3.31	2.24 / 3.09	2.15 / 2.93	2.08 / 2.79	2.02 / 2.70	1.98 / 2.61	1.94 / 2.54	1.90 / 2.47	1.85 / 2.37	1.80 / 2.30	1.73 / 2.18	1.68 / 2.09	1.63 / 2.00	1.57 / 1.90	1.54 / 1.84	1.49 / 1.76	1.46 / 1.71	1.42 / 1.64	1.39 / 1.60	1.37 / 1.56
70	3.98 / 7.01	3.13 / 4.92	2.74 / 4.08	2.50 / 3.60	2.35 / 3.29	2.23 / 3.07	2.14 / 2.91	2.07 / 2.77	2.01 / 2.67	1.97 / 2.59	1.93 / 2.51	1.89 / 2.45	1.84 / 2.35	1.79 / 2.28	1.72 / 2.15	1.67 / 2.07	1.62 / 1.98	1.56 / 1.88	1.53 / 1.82	1.47 / 1.74	1.45 / 1.69	1.40 / 1.62	1.37 / 1.56	1.35 / 1.53
80	3.96 / 6.96	3.11 / 4.88	2.72 / 4.04	2.48 / 3.56	2.33 / 3.25	2.21 / 3.04	2.12 / 2.87	2.05 / 2.74	1.99 / 2.64	1.95 / 2.55	1.91 / 2.48	1.88 / 2.41	1.82 / 2.32	1.77 / 2.24	1.70 / 2.11	1.65 / 2.03	1.60 / 1.94	1.54 / 1.84	1.51 / 1.78	1.45 / 1.70	1.42 / 1.65	1.38 / 1.57	1.35 / 1.52	1.32 / 1.49

Degrees of freedom for denominator

Table C. Critical values of F 141

Table C (continued) — Critical values of F

Degrees of freedom for denominator (lower margin): 100, 125, 150, 200, 400, 1000, ∞

	100	125	150	200	400	1000	∞
	1.28 / 1.43	1.25 / 1.37	1.22 / 1.33	1.19 / 1.28	1.13 / 1.19	1.08 / 1.11	1.00 / 1.00
	1.30 / 1.46	1.27 / 1.40	1.25 / 1.37	1.22 / 1.33	1.16 / 1.24	1.13 / 1.19	1.11 / 1.15
	1.34 / 1.51	1.31 / 1.46	1.29 / 1.43	1.26 / 1.39	1.22 / 1.32	1.19 / 1.28	1.17 / 1.25
	1.39 / 1.59	1.36 / 1.54	1.34 / 1.51	1.32 / 1.48	1.28 / 1.42	1.26 / 1.38	1.24 / 1.36
	1.42 / 1.64	1.39 / 1.59	1.37 / 1.56	1.35 / 1.53	1.32 / 1.47	1.30 / 1.44	1.28 / 1.41
	1.48 / 1.73	1.45 / 1.68	1.44 / 1.66	1.42 / 1.62	1.38 / 1.57	1.36 / 1.54	1.35 / 1.52
	1.51 / 1.79	1.49 / 1.75	1.47 / 1.72	1.45 / 1.69	1.42 / 1.64	1.41 / 1.61	1.40 / 1.59
	1.57 / 1.89	1.55 / 1.85	1.54 / 1.83	1.52 / 1.79	1.49 / 1.74	1.47 / 1.71	1.46 / 1.69
	1.63 / 1.98	1.60 / 1.94	1.59 / 1.91	1.57 / 1.88	1.54 / 1.84	1.53 / 1.81	1.52 / 1.79
	1.68 / 2.06	1.65 / 2.03	1.64 / 2.00	1.62 / 1.97	1.60 / 1.92	1.58 / 1.89	1.57 / 1.87
	1.75 / 2.19	1.72 / 2.15	1.71 / 2.12	1.69 / 2.09	1.67 / 2.04	1.65 / 2.01	1.64 / 1.99
	1.79 / 2.26	1.77 / 2.23	1.76 / 2.20	1.74 / 2.17	1.72 / 2.12	1.70 / 2.09	1.69 / 2.07
	1.35 / 2.36	1.83 / 2.33	1.82 / 2.30	1.80 / 2.28	1.78 / 2.23	1.76 / 2.20	1.75 / 2.18
	1.88 / 2.43	1.86 / 2.40	1.85 / 2.37	1.83 / 2.34	1.81 / 2.29	1.80 / 2.26	1.79 / 2.24
	1.92 / 2.51	1.90 / 2.47	1.89 / 2.44	1.87 / 2.41	1.85 / 2.37	1.84 / 2.34	1.83 / 2.32
	1.97 / 2.59	1.95 / 2.56	1.94 / 2.53	1.92 / 2.50	1.90 / 2.46	1.89 / 2.43	1.88 / 2.41
	2.03 / 2.69	2.01 / 2.65	2.00 / 2.62	1.98 / 2.60	1.96 / 2.55	1.95 / 2.53	1.94 / 2.51
	2.10 / 2.82	2.08 / 2.79	2.07 / 2.76	2.05 / 2.73	2.03 / 2.69	2.02 / 2.66	2.01 / 2.64
	2.19 / 2.99	2.17 / 2.95	2.16 / 2.92	2.14 / 2.90	2.12 / 2.85	2.10 / 2.82	2.09 / 2.80
	2.30 / 3.20	2.29 / 3.17	2.27 / 3.13	2.26 / 3.11	2.23 / 3.06	2.22 / 3.04	2.21 / 3.02
	2.46 / 3.51	2.44 / 3.47	2.43 / 3.44	2.41 / 3.41	2.39 / 3.36	2.38 / 3.34	2.37 / 3.32
	2.70 / 3.98	2.68 / 3.94	2.67 / 3.91	2.65 / 3.88	2.62 / 3.83	2.61 / 3.80	2.60 / 3.78
	3.09 / 4.82	3.07 / 4.78	3.06 / 4.75	3.04 / 4.71	3.02 / 4.66	3.00 / 4.62	2.99 / 4.60
	3.94 / 6.90	3.92 / 6.84	3.91 / 6.81	3.89 / 6.76	3.86 / 6.70	3.85 / 6.66	3.84 / 6.64

Table C₁ Values of F exceeded by 0.025 of the values in the sampling distribution

If, in testing the homogeneity of two sample variances, the larger variance is placed over the smaller, the number of ratios greater than any given value, equal to or greater than unity, is doubled. Therefore use of the values tabulated below, in comparing two sample variances, will provide a 0.05 level of significance.

		df for larger variance (numerator)									
		4	5	6	7	8	9	10	12	15	20
	4	9.60	9.36	9.20	9.07	8.98	8.90	8.84	8.75	8.66	8.56
	5	7.39	7.15	6.98	6.85	6.76	6.68	6.62	6.52	6.43	6.33
df	6	6.23	5.99	5.82	5.70	5.60	5.52	5.46	5.37	5.27	5.17
for	7	5.52	5.29	5.12	4.99	4.90	4.82	4.76	4.67	4.57	4.47
smaller	8	5.05	4.82	4.65	4.53	4.43	4.36	4.30	4.20	4.10	4.00
variance	9	4.72	4.48	4.32	4.20	4.10	4.03	3.96	3.87	3.77	3.67
(denomi-	10	4.47	4.24	4.07	3.95	3.85	3.78	3.72	3.62	3.52	3.42
nator)	12	4.12	3.89	3.73	3.61	3.51	3.44	3.37	3.28	3.18	3.07
	15	3.80	3.58	3.41	3.29	3.20	3.12	3.06	2.96	2.86	2.76
	20	3.51	3.29	3.13	3.01	2.91	2.84	2.77	2.68	2.57	2.46

Interpolation may be performed using reciprocals of the degrees of freedom.

Table D Critical values of *A*

For any given value of n − 1, the table shows the values of A corresponding to various levels of probability. A is significant at a given level if it is equal to or <u>less than</u> the value shown in the table.

n − 1*	Level of significance for one-tailed test					n − 1*
	.05	.025	.01	.005	.0005	
	Level of significance for two-tailed test					
	.10	.05	.02	.01	.001	
1	0.5125	0.5031	0.50049	0.50012	0.5000012	1
2	0.412	0.369	0.347	0.340	0.334	2
3	0.385	0.324	0.286	0.272	0.254	3
4	0.376	0.304	0.257	0.238	0.211	4
5	0.372	0.293	0.240	0.218	0.184	5
6	0.370	0.286	0.230	0.205	0.167	6
7	0.369	0.281	0.222	0.196	0.155	7
8	0.368	0.278	0.217	0.190	0.146	8
9	0.368	0.276	0.213	0.185	0.139	9
10	0.368	0.274	0.210	0.181	0.134	10
11	0.368	0.273	0.207	0.178	0.130	11
12	0.368	0.271	0.205	0.176	0.126	12
13	0.368	0.270	0.204	0.174	0.124	13
14	0.368	0.270	0.202	0.172	0.121	14
15	0.368	0.269	0.201	0.170	0.119	15
16	0.368	0.268	0.200	0.169	0.117	16
17	0.368	0.268	0.199	0.168	0.116	17
18	0.368	0.267	0.198	0.167	0.114	18
19	0.368	0.267	0.197	0.166	0.113	19
20	0.368	0.266	0.197	0.165	0.112	20
21	0.368	0.266	0.196	0.165	0.111	21
22	0.368	0.266	0.196	0.164	0.110	22
23	0.368	0.266	0.195	0.163	0.109	23
24	0.368	0.265	0.195	0.163	0.108	24
25	0.368	0.265	0.194	0.162	0.108	25
26	0.368	0.265	0.194	0.162	0.107	26
27	0.368	0.265	0.193	0.161	0.107	27
28	0.368	0.265	0.193	0.161	0.106	28
29	0.368	0.264	0.193	0.161	0.106	29
30	0.368	0.264	0.193	0.160	0.105	30
40	0.368	0.263	0.191	0.158	0.102	40
60	0.369	0.262	0.189	0.155	0.099	60
120	0.369	0.261	0.187	0.153	0.095	120
∞	0.370	0.260	0.185	0.151	0.092	∞

*n = number of pairs

Table E   Percentage points of the Studentized range

Error df	α	k = number of means or number of steps between ordered means									
		2	3	4	5	6	7	8	9	10	11
5	.05	3.64	4.60	5.22	5.67	6.03	6.33	6.58	6.80	6.99	7.17
	.01	5.70	6.98	7.80	8.42	8.91	9.32	9.67	9.97	10.24	10.48
6	.05	3.46	4.34	4.90	5.30	5.63	5.90	6.12	6.32	6.49	6.65
	.01	5.24	6.33	7.03	7.56	7.97	8.32	8.61	8.87	9.10	9.30
7	.05	3.34	4.16	4.68	5.06	5.36	5.61	5.82	6.00	6.16	6.30
	.01	4.95	5.92	6.54	7.01	7.37	7.68	7.94	8.17	8.37	8.55
8	.05	3.26	4.04	4.53	4.89	5.17	5.40	5.60	5.77	5.92	6.05
	.01	4.75	5.64	6.20	6.62	6.96	7.24	7.47	7.68	7.86	8.03
9	.05	3.20	3.95	4.41	4.76	5.02	5.24	5.43	5.59	5.74	5.87
	.01	4.60	5.43	5.96	6.35	6.66	6.91	7.13	7.33	7.49	7.65
10	.05	3.15	3.88	4.33	4.65	4.91	5.12	5.30	5.46	5.60	5.72
	.01	4.48	5.27	5.77	6.14	6.43	6.67	6.87	7.05	7.21	7.36
11	.05	3.11	3.82	4.26	4.57	4.82	5.03	5.20	5.35	5.49	5.61
	.01	4.39	5.15	5.62	5.97	6.25	6.48	6.67	6.84	6.99	7.13
12	.05	3.08	3.77	4.20	4.51	4.75	4.95	5.12	5.27	5.39	5.51
	.01	4.32	5.05	5.50	5.84	6.10	6.32	6.51	6.67	6.81	6.94
13	.05	3.06	3.73	4.15	4.45	4.69	4.88	5.05	5.19	5.32	5.43
	.01	4.26	4.96	5.40	5.73	5.98	6.19	6.37	6.53	6.67	6.79
14	.05	3.03	3.70	4.11	4.41	4.64	4.83	4.99	5.13	5.25	5.36
	.01	4.21	4.89	5.32	5.63	5.88	6.08	6.26	6.41	6.54	6.66
15	.05	3.01	3.67	4.08	4.37	4.59	4.78	4.94	5.08	5.20	5.31
	.01	4.17	4.84	5.25	5.56	5.80	5.99	6.16	6.31	6.44	6.55
16	.05	3.00	3.65	4.05	4.33	4.56	4.74	4.90	5.03	5.15	5.26
	.01	4.13	4.79	5.19	5.49	5.72	5.92	6.08	6.22	6.35	6.46
17	.05	2.98	3.63	4.02	4.30	4.52	4.70	4.86	4.99	5.11	5.21
	.01	4.10	4.74	5.14	5.43	5.66	5.85	6.01	6.15	6.27	6.38
18	.05	2.97	3.61	4.00	4.28	4.49	4.67	4.82	4.96	5.07	5.17
	.01	4.07	4.70	5.09	5.38	5.60	5.79	5.94	6.08	6.20	6.31
19	.05	2.96	3.59	3.98	4.25	4.47	4.65	4.79	4.92	5.04	5.14
	.01	4.05	4.67	5.05	5.33	5.55	5.73	5.89	6.02	6.14	6.25
20	.05	2.95	3.58	3.96	4.23	4.45	4.62	4.77	4.90	5.01	5.11
	.01	4.02	4.64	5.02	5.29	5.51	5.69	5.84	5.97	6.09	6.19
24	.05	2.92	3.53	3.90	4.17	4.37	4.54	4.68	4.81	4.92	5.01
	.01	3.96	4.55	4.91	5.17	5.37	5.54	5.69	5.81	5.92	6.02
30	.05	2.89	3.49	3.85	4.10	4.30	4.46	4.60	4.72	4.82	4.92
	.01	3.89	4.45	4.80	5.05	5.24	5.40	5.54	5.65	5.76	5.85
40	.05	2.86	3.44	3.79	4.04	4.23	4.39	4.52	4.63	4.73	4.82
	.01	3.82	4.37	4.70	4.93	5.11	5.26	5.39	5.50	5.60	5.69
60	.05	2.83	3.40	3.74	3.98	4.16	4.31	4.44	4.55	4.65	4.73
	.01	3.76	4.28	4.59	4.82	4.99	5.13	5.25	5.36	5.45	5.53
120	.05	2.80	3.36	3.68	3.92	4.10	4.24	4.36	4.47	4.56	4.64
	.01	3.70	4.20	4.50	4.71	4.87	5.01	5.12	5.21	5.30	5.37
∞	.05	2.77	3.31	3.63	3.86	4.03	4.17	4.29	4.39	4.47	4.55
	.01	3.64	4.12	4.40	4.60	4.76	4.88	4.99	5.08	5.16	5.23

Table F   Transformation of r to z_r

r	z_r	r	z_r	r	z_r
.01	.010	.34	.354	.67	.811
.02	.020	.35	.366	.68	.829
.03	.030	.36	.377	.69	.848
.04	.040	.37	.389	.70	.867
.05	.050	.38	.400	.71	.887
.06	.060	.39	.412	.72	.908
.07	.070	.40	.424	.73	.929
.08	.080	.41	.436	.74	.950
.09	.090	.42	.448	.75	.973
.10	.100	.43	.460	.76	.996
.11	.110	.44	.472	.77	1.020
.12	.121	.45	.485	.78	1.045
.13	.131	.46	.497	.79	1.071
.14	.141	.47	.510	.80	1.099
.15	.151	.48	.523	.81	1.127
.16	.161	.49	.536	.82	1.157
.17	.172	.50	.549	.83	1.188
.18	.181	.51	.563	.84	1.221
.19	.192	.52	.577	.85	1.256
.20	.203	.53	.590	.86	1.293
.21	.214	.54	.604	.87	1.333
.22	.224	.55	.618	.88	1.376
.23	.234	.56	.633	.89	1.422
.24	.245	.57	.648	.90	1.472
.25	.256	.58	.663	.91	1.528
.26	.266	.59	.678	.92	1.589
.27	.277	.60	.693	.93	1.658
.28	.288	.61	.709	.94	1.738
.29	.299	.62	.725	.95	1.832
.30	.309	.63	.741	.96	1.946
.31	.321	.64	.758	.97	2.092
.32	.332	.65	.775	.98	2.298
.33	.343	.66	.793	.99	2.647

Table G   Functions of r

r	$\sqrt{r}$	r^2	$\sqrt{r-r^2}$	$\sqrt{1-r}$	$1-r^2$	$\sqrt{1-r^2}$	$100(1-k)$	r
						k	% Eff.	
1.00	1.0000	1.0000	0.0000	0.0000	0.0000	0.0000	100.00	1.00
.99	.9950	.9801	.0995	.1000	.0199	.1411	85.89	.99
.98	.9899	.9604	.1400	.1414	.0396	.1990	80.10	.98
.97	.9849	.9409	.1706	.1732	.0591	.2431	75.69	.97
.96	.9798	.9216	.1960	.2000	.0784	.2800	72.00	.96
.95	.9747	.9025	.2179	.2236	.0975	.3122	68.78	.95
.94	.9695	.8836	.2375	.2449	.1164	.3412	65.88	.94
.93	.9644	.8649	.2551	.2646	.1351	.3676	63.24	.93
.92	.9592	.8464	.2713	.2828	.1536	.3919	60.81	.92
.91	.9539	.8281	.2862	.3000	.1719	.4146	58.54	.91
.90	.9487	.8100	.3000	.3162	.1900	.4359	56.41	.90
.89	.9434	.7921	.3129	.3317	.2079	.4560	54.40	.89
.88	.9381	.7744	.3250	.3464	.2256	.4750	52.50	.88
.87	.9327	.7569	.3363	.3606	.2431	.4931	50.69	.87
.86	.9274	.7396	.3470	.3742	.2604	.5103	48.97	.86
.85	.9220	.7225	.3571	.3873	.2775	.5268	47.32	.85
.84	.9165	.7056	.3666	.4000	.2944	.5426	45.74	.84
.83	.9110	.6889	.3756	.4123	.3111	.5578	44.22	.83
.82	.9055	.6724	.3842	.4243	.3276	.5724	42.76	.82
.81	.9000	.6561	.3923	.4359	.3439	.5864	41.36	.81
.80	.8944	.6400	.4000	.4472	.3600	.6000	40.00	.80
.79	.8888	.6241	.4073	.4583	.3759	.6131	38.69	.79
.78	.8832	.6084	.4142	.4690	.3916	.6258	37.42	.78
.77	.8775	.5929	.4208	.4796	.4071	.6380	36.20	.77
.76	.8718	.5776	.4271	.4899	.4224	.6499	35.01	.76
.75	.8660	.5625	.4330	.5000	.4375	.6614	33.86	.75
.74	.8602	.5476	.4386	.5099	.4524	.6726	32.74	.74
.73	.8544	.5329	.4440	.5196	.4671	.6834	31.66	.73
.72	.8485	.5184	.4490	.5292	.4816	.6940	30.60	.72
.71	.8426	.5041	.4538	.5385	.4959	.7042	29.58	.71
.70	.8367	.4900	.4583	.5477	.5100	.7141	28.59	.70
.69	.8307	.4761	.4625	.5568	.5239	.7238	27.62	.69
.68	.8246	.4624	.4665	.5657	.5376	.7332	26.68	.68
.67	.8185	.4489	.4702	.5745	.5511	.7424	25.76	.67
.66	.8124	.4356	.4737	.5831	.5644	.7513	24.87	.66
.65	.8062	.4225	.4770	.5916	.5775	.7599	24.01	.65
.64	.8000	.4096	.4800	.6000	.5904	.7684	23.16	.64
.63	.7937	.3969	.4828	.6083	.6031	.7766	22.34	.63
.62	.7874	.3844	.4854	.6164	.6156	.7846	21.54	.62
.61	.7810	.3721	.4877	.6245	.6279	.7924	20.76	.61
.60	.7746	.3600	.4899	.6325	.6400	.8000	20.00	.60
.59	.7681	.3481	.4918	.6403	.6519	.8074	19.26	.59
.58	.7616	.3364	.4936	.6481	.6636	.8146	18.54	.58
.57	.7550	.3249	.4951	.6557	.6751	.8216	17.84	.57
.56	.7483	.3136	.4964	.6633	.6864	.8285	17.15	.56
.55	.7416	.3025	.4975	.6708	.6975	.8352	16.48	.55
.54	.7348	.2916	.4984	.6782	.7084	.8417	15.83	.54
.53	.7280	.2809	.4991	.6856	.7191	.8480	15.20	.53
.52	.7211	.2704	.4996	.6928	.7296	.8542	14.58	.52
.51	.7141	.2601	.4999	.7000	.7399	.8602	13.98	.51
.50	.7071	.2500	.5000	.7071	.7500	.8660	13.40	.50

Table G (*Continued*)

r	$\sqrt{r}$	r^2	$\sqrt{r-r^2}$	$\sqrt{1-r}$	$1-r^2$	$\sqrt{1-r^2}$	$100(1-k)$	r
						k	% Eff.	
.50	.7071	.2500	.5000	.7071	.7500	.8660	13.40	.50
.49	.7000	.2401	.4999	.7141	.7599	.8717	12.83	.49
.48	.6928	.2304	.4996	.7211	.7696	.8773	12.27	.48
.47	.6856	.2209	.4991	.7280	.7791	.8827	11.73	.47
.46	.6782	.2116	.4984	.7348	.7884	.8879	11.21	.46
.45	.6708	.2025	.4975	.7416	.7975	.8930	10.70	.45
.44	.6633	.1936	.4964	.7483	.8064	.8980	10.20	.44
.43	.6557	.1849	.4951	.7550	.8151	.9028	9.72	.43
.42	.6481	.1764	.4936	.7616	.8236	.9075	9.25	.42
.41	.6403	.1681	.4918	.7681	.8319	.9121	8.79	.41
.40	.6325	.1600	.4899	.7746	.8400	.9165	8.35	.40
.39	.6245	.1521	.4877	.7810	.8479	.9208	7.92	.39
.38	.6164	.1444	.4854	.7874	.8556	.9250	7.50	.38
.37	.6083	.1369	.4828	.7937	.8631	.9290	7.10	.37
.36	.6000	.1296	.4800	.8000	.8704	.9330	6.70	.36
.35	.5916	.1225	.4770	.8062	.8775	.9367	6.33	.35
.34	.5831	.1156	.4737	.8124	.8844	.9404	5.96	.34
.33	.5745	.1089	.4702	.8185	.8911	.9440	5.60	.33
.32	.5657	.1024	.4665	.8246	.8976	.9474	5.25	.32
.31	.5568	.0961	.4625	.8307	.9039	.9507	4.93	.31
.30	.5477	.0900	.4583	.8367	.9100	.9539	4.61	.30
.29	.5385	.0841	.4538	.8426	.9159	.9570	4.30	.29
.28	.5292	.0784	.4490	.8485	.9216	.9600	4.00	.28
.27	.5196	.0729	.4440	.8544	.9271	.9629	3.71	.27
.26	.5099	.0676	.4386	.8602	.9324	.9656	3.44	.26
.25	.5000	.0625	.4330	.8660	.9375	.9682	3.18	.25
.24	.4899	.0576	.4271	.8718	.9424	.9708	2.92	.24
.23	.4796	.0529	.4208	.8775	.9471	.9732	2.68	.23
.22	.4690	.0484	.4142	.8832	.9516	.9755	2.45	.22
.21	.4583	.0441	.4073	.8888	.9559	.9777	2.23	.21
.20	.4472	.0400	.4000	.8944	.9600	.9798	2.02	.20
.19	.4359	.0361	.3923	.9000	.9639	.9818	1.82	.19
.18	.4243	.0324	.3842	.9055	.9676	.9837	1.63	.18
.17	.4123	.0289	.3756	.9110	.9711	.9854	1.46	.17
.16	.4000	.0256	.3666	.9165	.9744	.9871	1.29	.16
.15	.3873	.0225	.3571	.9220	.9775	.9887	1.13	.15
.14	.3742	.0196	.3470	.9274	.9804	.9902	.98	.14
.13	.3606	.0169	.3363	.9327	.9831	.9915	.85	.13
.12	.3464	.0144	.3250	.9381	.9856	.9928	.72	.12
.11	.3317	.0121	.3129	.9434	.9879	.9939	.61	.11
.10	.3162	.0100	.3000	.9487	.9900	.9950	.50	.10
.09	.3000	.0081	.2862	.9539	.9919	.9959	.41	.09
.08	.2828	.0064	.2713	.9592	.9936	.9968	.32	.08
.07	.2646	.0049	.2551	.9644	.9951	.9975	.25	.07
.06	.2449	.0036	.2375	.9695	.9964	.9982	.18	.06
.05	.2236	.0025	.2179	.9747	.9975	.9987	.13	.05
.04	.2000	.0016	.1960	.9798	.9984	.9992	.08	.04
.03	.1732	.0009	.1706	.9849	.9991	.9995	.05	.03
.02	.1414	.0004	.1400	.9899	.9996	.9998	.02	.02
.01	.1000	.0001	.0995	.9950	.9999	.9999	.01	.01
.00	.0000	.0000	.0000	1.0000	1.0000	1.0000	.00	.00

Answers

CHAPTER 1 EXERCISES

1. If $p = 0.05$, the chances are 5 in 100 or 1 in 20 that the event will occur.

2. If $p = 0.01$, the chances are 1 in 100 that the event will occur.

3. $p = 0.40$

4. $p = 0.75$

5. 1/52

6. 4/52 or 1/13

7. 13/52 or 1/4

8. 26/52 or 1/2

9. 2/52 or 1/26

10. 16/52 or 4/13

11. 0.0630

12. 0.9370

13. 0.9591

14. 0.0409

15. 0.5000

16. 0.5000

17. 0.0043

18. 0.0202

19. 0.0250

20. 0.0250

21. 0.0099

22. 0.0099

23. 0.9901

24. 0.9901

25. $p = 0.0708$

26. $p = 0.7673$

27. $p = 0.2327$

28. $p = 0.4207$

29. $p = 0.5793$

30. $p = 0.1616$

31. $p = 0.2628$

32. $p = 0.0214$

33. $p = 0.2714$

34. $p = 0.0214$

35. $p = 0.0072$

36.

Second draw	First draw						
	1	**2**	**3**	**4**	**5**	**6**	**7**
1	1.0	1.5	2.0	2.5	3.0	3.5	4.0
2	1.5	2.0	2.5	3.0	3.5	4.0	4.5
3	2.0	2.5	3.0	3.5	4.0	4.5	5.0
4	2.5	3.0	3.5	4.0	4.5	5.0	5.5
5	3.0	3.5	4.0	4.5	5.0	5.5	6.0
6	3.5	4.0	4.5	5.0	5.5	6.0	6.5
7	4.0	4.5	5.0	5.5	6.0	6.5	7.0

37.

$\bar{X}$	f
7.0	1
6.5	2
6.0	3
5.5	4
5.0	5
4.5	6
4.0	7
3.5	6
3.0	5
2.5	4
2.0	3
1.5	2
1.0	1
	$N_{\bar{X}} = 49$

38.

$\bar{X}$	f	$p_{\bar{X}}$
7.0	1	0.0204
6.5	2	0.0408
6.0	3	0.0612
5.5	4	0.0816
5.0	5	0.1020
4.5	6	0.1224
4.0	7	0.1429
3.5	6	0.1224
3.0	5	0.1020
2.5	4	0.0816
2.0	3	0.0612
1.5	2	0.0408
1.0	1	0.0204
	$N_{\bar{X}} = 49$	$\sum p_{\bar{X}} = 0.9997$

39. $p = 0.0204$

40. $p = 0.0408 + 0.0204 = 0.0612$

41. $p = 2(0.0612) = 0.1224$

42. $p = 2(0.0612 + 0.0408 + 0.0204)$
    $= 0.2448$

43. $p = 0.1020 + 0.0816 + 0.0612$
    $+ 0.0408 + 0.0204$
    $= 0.3060$

CHAPTER 1 TEST

1. b   2. c   3. d   4. a

5. c   6. a   7. d   8. a

9. c   10. b   11. a   12. c

13. a   14. c   15. d

16. a) $z = 1.67, p = 0.0475$ b) $z = -1.33, p = 0.0918$

   c) $z = 2.33, p = 0.0099$ d) $z = -3.00, p = 0.0013$

   e) $z = -2.67, p = 0.0038$ f) $z = 0.33, p = 0.3707$

   g) $z = 3.00, p = 0.0013$ h) $z = -0.67, p = 0.2514$

   i) $z = 1.67, p = 0.0950$ j) $z = -2.00, p = 0.0456$

   k) $z = -3.00, p = 0.0026$ l) $z = 3.00, p = 0.0026$

   m) $z = 0.67, p = 0.5028$

17.

$\bar{X}$	f	$p_{\bar{X}}$
8.0	1	0.0123
7.5	2	0.0247
7.0	3	0.0370
6.5	4	0.0494
6.0	5	0.0617
5.5	6	0.0741
5.0	7	0.0864
4.5	8	0.0988
4.0	9	0.1111
3.5	8	0.0988
3.0	7	0.0864
2.5	6	0.0741
2.0	5	0.0617
1.5	4	0.0494
1.0	3	0.0370
0.5	2	0.0247
0.0	1	0.0123
$N_{\bar{X}} = 81$	$\sum p_{\bar{X}} = 0.9999$	

a) $p = 0.037$

b) $p = 0.0246$

c) $p = 0.1234$

d) $p = 0.0740$

e) $p = 0.1851$

CHAPTER 2 EXERCISES

1. Since the two-tailed p-value is 0.0408 and this is less than 0.05, we make the reject decision. It is not likely that the sample was drawn from the indicated population.

2. Since the one-tailed p-value is 0.0204, we make the reject decision. It is not likely that the sample was drawn from the indicated population.

3. Since the two-tailed p-value is 0.1224 (0.0204 + 0.0408 + 0.0408 + 0.0204), we make the fail to reject decision.

4. Since the two-tailed p-value is 0.0408, we make the reject decision.

5. Since the one-tailed p-value of 0.0612 is greater than 0.05, we make the fail to reject decision.

6. Since the one-tailed p-value of 5.5 or greater is 0.2040, and this is greater than $p = 0.05$, we make the fail to reject decision.

7. Fail to reject   8. Reject

9. Fail to reject   10. Reject

11. Fail to reject   12. Fail to reject

13. Reject   14. Fail to reject

15. $\alpha = 0.01$ 16. $\alpha = 0.05$

17. H_0: The number of hours of study has no effect upon academic performance.

18. H_0: The type of dial has no effect on speed-of-reading scores.

19. H_1: The number of hours of study has an effect upon academic performance.

20. H_1: The type of dial has an effect on speed-of-reading scores.

21. H_0: The mean of the 8-hour group will be equal to or less than the mean of the 4-hour group, i.e., $\mu_8 \leq \mu_4$.
    H_1: The mean of the 8-hour group will be greater than the mean of the 4-hour group, i.e., $\mu_8 > \mu_4$.

22. H_0: The mean speed of reading a rectangular dial is equal to or less than the mean speed of reading the circular dial, i.e., $\mu_r \leq \mu_c$.
    H_1: The mean speed of reading a rectangular dial is greater than the mean speed of reading a circular dial, i.e., $\mu_r > \mu_c$.

23. Since two-tailed $p = 0.014$ and this is greater than α, we fail to reject H_0.

24. Since one-tailed $p = 0.009$ is less than α, we reject H_0 and assert H_1.

25. Since two-tailed $p = 0.06$ is greater than α, we fail to reject H_0.

26. Since two-tailed $p = 0.04$ is less than α, we reject H_0 and assert H_1.

27. Since one-tailed p-value is equal to α, we reject H_0 and assert H_1.

28. Since two-tailed $p = 0.04$ is greater than α, we fail to reject H_0.

29. Since one-tailed $p = 0.06$ is greater than α, we fail to reject H_0.

30. Since two-tailed $p = 0.008$ is less than α, we reject H_0 and assert H_1.

31. $z \leq -2.58$ and $z \geq 2.58$.

32. $z \geq 2.33$ 33. $z \leq -1.65$

34. z lies in the critical region. Reject H_0.

35. z lies in the critical region. Reject H_0.

36. z lies in the critical region. Reject H_0.

37. z does not lie in the critical region. Do not reject H_0.

38. z does not lie in the critical region. Do not reject H_0.

39. z does not lie in the critical region. Do not reject H_0.

40. z does not lie in the critical region. Do not reject H_0.

CHAPTER 2 TEST

1. a   2. a   3. b   4. d

5. b   6. c   7. a   8. c

9. b   10. d

11.

$\bar{X}$	f	$p_{\bar{x}}$
0.0	1	0.0059
0.5	2	0.0118
1.0	3	0.0178
1.5	4	0.0237
2.0	5	0.0296
2.5	6	0.0355
3.0	7	0.0414
3.5	8	0.0473
4.0	9	0.0533
4.5	10	0.0592
5.0	11	0.0651
5.5	12	0.0710
6.0	13	0.0769
6.5	12	0.0710
7.0	11	0.0651
7.5	10	0.0592
8.0	9	0.0533
8.5	8	0.0473
9.0	7	0.0414
9.5	6	0.0355
10.0	5	0.0296
10.5	4	0.0237
11.0	3	0.0178
11.5	2	0.0118
12.0	1	0.0059

12. a) $p = 0.0355$, reject H_0 and assert H_1.

    b) $p = 0.1184$, fail to reject H_0.

    c) $p = 0.710$, fail to reject H_0.

    d) $p = 0.0592$, fail to reject H_0.

    e) $p = 0.0354$, reject H_0 and assert H_1.

    f) $p = 0.0354$, fail to reject H_0.

    g) $p = 0.0177$, fail to reject H_0.

CHAPTER 3 EXERCISES

1. $z = 1.67$

2. $z = -1.73$

3. $z = -0.80$

4. $z = 2.22$

5. $z = -1.70$

6. $\sigma_{\bar{x}} = 5.06$

7. $\sigma_{\bar{x}} = 1.60$

8. $\sigma_{\bar{x}} = 0.51$

9. $\sigma_{\bar{x}} = 20.00$

10. $\sigma_{\bar{x}} = 4.00$

11. 1) *Null hypothesis* (H_0): The mean of the population from which the sample was drawn is 70, that is, $\mu = \mu_0 = 70$.

 2) *Alternative hypothesis* (H_1): The mean of the population from which the sample was drawn is not equal to 70, that is, $\mu \neq 70$.

 3) *Significance level*: $\alpha = 0.05$. If the difference between the sample mean and the specified population mean is so extreme that its associated probability of occurrence under H_0 is equal to or less than 0.05, we shall reject H_0.

 4) *Sampling distribution*: Since the population mean and standard deviation are known, the standard normal curve is the appropriate sampling distribution.

 5) *Critical region for rejection of H_0*: In Table A, we find that ± 1.96 includes 5% of the area beyond it. Thus, if obtained z is equal to or greater than 1.96 or equal to or less than -1.96, we shall reject H_0.

 The steps in the test of significance are as follows:

 Step 1. $\sigma_{\bar{x}} = \dfrac{15}{\sqrt{40}} = 2.37$

 Step 2. $z = \dfrac{60 - 70}{2.37} = -4.22$

 Conclusion: Since $z = -4.22$ is in the critical region, we reject H_0 and assert H_1.

12. $\sum X = 112, \sum X^2 = 2316, N = 6$

 $\sum x^2 = 2316 - \dfrac{(112)^2}{6} = 225.33$

 $s_{\bar{x}} = \sqrt{\dfrac{225.33}{(6)(5)}} = 2.74$

13. $\sum X = 336, \sum X^2 = 14{,}980, N = 8$

 $\sum x^2 = 14{,}980 - \dfrac{(336)^2}{8} = 868.00$

 $s_{\bar{x}} = \sqrt{\dfrac{868.00}{(8)(7)}} = 3.94$

14. $\sum X = 90, \sum X^2 = 930, N = 10$

 $\sum x^2 = 930 - \dfrac{(90)^2}{10} = 120.00$

 $s_{\bar{x}} = \sqrt{\dfrac{120.00}{(10)(9)}} = 1.15$

15. $\sum X = 212, \sum X^2 = 6036, N = 11$

 $\sum X^2 = 6036 - \dfrac{(212)^2}{11} = 1950.18$

 $s_{\bar{x}} = \sqrt{\dfrac{1950.18}{(11)(10)}} = 4.21$

16. $t = -2.50$

17. $t = 1.77$

18. $t = 2.30$

19. $t = -2.24$

20. df $= 17$

21. df $= 1$

22. df $= 91$

23. df $= 107$

24. df $= 99$

25. df $= 10$

26. df $= 41$

27. df $= 30$

28. df $= 2$

29. df $= 71$

30. Since the critical value at $\alpha = 0.01$ with 30 df, two-tailed test, is 2.750, obtained $t = 2.48$ is not in the critical region. We fail to reject H_0.

31. Since the critical value at $\alpha = 0.01$ with 60 df, one-tailed test, is 2.390, obtained $t = 2.44$ is in the critical region. We reject H_0 and assert H_1.

32. Since the critical value at $\alpha = 0.05$ with 11 df, two-tailed test, is 2.201, obtained $t = 2.25$ is in the critical region. We reject H_0 and assert H_1.

33. 1) *Null hypothesis* (H_0): $\mu = \mu_0$
    2) *Alternative hypothesis* (H_1): $\mu \neq \mu_0$
    3) *Statistical test*: Since the comparison involves a sample mean and a hypothetical population value and α is unknown, the Student t-ratio, one-sample case, is appropriate.
    4) *Significance level*: $\alpha = 0.05$
    5) *Sampling distribution*: The sampling distribution is the Student t-distribution with df $= N - 1 = 14 - 1 = 13$.
    6) *Critical region*: By referring to Table B, we find that the critical value for significance at $\alpha = 0.05$, two-tailed test, when df $= 13$ is 2.160.

    Step 1. $\sum X = 164$, $N = 14$,

    $$\bar{X} = \frac{164}{14} = 11.71$$

    Step 2. $\sum x^2 = 2016 - \frac{(164)^2}{14} = 94.86$

    Step 3. $s_{\bar{X}} = \sqrt{\frac{94.86}{182}} = 0.72$

    Step 4. $t = \frac{11.71 - 10}{0.72} = \frac{1.71}{0.72} = 2.375$

    Conclusion: Since obtained t is in the critical region, we reject H_0 and assert H_1.

34.

X	D $(X - \mu_0)$	D^2
8	-2	4
12	2	4
15	5	25
9	-1	1
13	3	9
11	1	1
17	7	49
7	-3	9
11	1	1
12	2	4
14	4	16
10	0	0
12	2	4
13	3	9
	$\sum D = 24$	$\sum D^2 = 136$

Step 1. $\sum D = 24$

Step 2. $\sum D^2 = 136$

Step 3. $A = \frac{136}{(24)^2} = 0.236$

Conclusion: Reference to Table D shows that $A \leq 0.270$ is required for significance at $\alpha = 0.05$, two-tailed test, when df $= 13$. Since obtained t is less than the critical value, it is in the critical region. Thus we reject H_0 and assert H_1.

35. $z = \frac{0.485 - 0}{\sqrt{\frac{1}{23}}} = \frac{0.485}{0.209} = 2.32$

36. $z = \frac{-1.157 - 0}{\sqrt{\frac{1}{7}}} = \frac{-1.157}{0.378} = -3.06$

37. $z = \frac{0.151 - 0}{\sqrt{\frac{1}{497}}} = \frac{0.151}{0.045} = 3.36$

38. $z = \frac{0.775 - 0.693}{\sqrt{\frac{1}{97}}} = \frac{0.082}{0.102} = 0.80$

39. Since the critical value at $\alpha = 0.05$, one-tailed test, is $z \geq 1.65$, obtained $z = 2.32$ is in the critical region. We reject H_0 and assert H_1.

40. Since the critical value at $\alpha = 0.01$, two-tailed test, is $z \geq 2.58$ or $z \leq -2.58$, obtained $z = -3.06$ is in the critical region. We reject H_0 and assert H_1.

41. Since the critical value at $\alpha = 0.05$, two-tailed test, is $z \geq 1.96$ or $z = \leq -1.96$, obtained $z = 3.36$ is in the critical region. We reject H_0 and assert H_1.

42. Since the critical value at $\alpha = 0.05$, one-tailed test, is $z \geq 1.65$, obtained $z = 0.80$ is not in the critical region. We fail to reject H_0.

CHAPTER 3 TEST

1. b	2. a	3. d	4. a
5. c	6. a	7. c	8. b
9. d	10. c	11. a	12. b
13. a	14. c	15. b	

16. $z = \dfrac{53 - 50}{\sigma_{\bar{X}}}$

$\sigma_{\bar{X}} = \dfrac{9}{\sqrt{20}} = 2.01$

Therefore $z = \dfrac{3}{2.01} = 1.49.$

Since the critical value of z for a nondirectional hypothesis at $\alpha = 0.01$ is ± 2.58, obtained z is not in the critical region. We fail to reject H_0.

17. $t = \dfrac{53 - 50}{s_{\bar{X}}}$

$\sum x^2 = 39,396 - \dfrac{(742)^2}{14} = 70$

$s_{\bar{X}} = \sqrt{\dfrac{\sum x^2}{N(N-1)}} = \sqrt{\dfrac{70}{(14)(13)}} = 0.62$

$t = \dfrac{3}{0.62} = 4.84$

18. $\sum X^2 = 12,845, \sum X = 391$

$n = 12, \text{df} = 11$

$\bar{X} = 32.58$

$\sum x^2 = 12,845 - \dfrac{(391)^2}{12} = 104.92$

$s_{\bar{X}} = \sqrt{\dfrac{104.92}{(12)(11)}} = 0.89$

$t = \dfrac{32.58 - 30.00}{0.89} = 2.90$

Since the critical value for the nondirectional hypothesis at $\alpha = 0.01$ consists of absolute values of t equal to or greater than 3.106, obtained t is not in the critical region. We fail to reject H_0.

$\sum D = 31, \sum D^2 = 185$

$A = \dfrac{185}{961} = 0.193$

Since the critical value of A consists of values of the test statistic equal to or less than 0.178, A is not in the critical region. We fail to reject H_0.

19. $z = \dfrac{0.618 - 0.203}{\sqrt{\dfrac{1}{57}}} = \dfrac{0.415}{0.132} = 3.14$

Since the critical value of z for the directional hypothesis at $\alpha = 0.01$ consists of values of the test statistic equal to or greater than 2.33, obtained z is in the critical region. We reject H_0 and assert H_1.

CHAPTER 4 EXERCISES

1. $\mu_2 - \mu_1 = 6.5$ or $\mu_2 - 6.5 = \mu_1$

2. $\mu_1 - \mu_2 = 9$ or $\mu_1 - 9 = \mu_2$

3. $\mu_A - \mu_B = 0$ or $\mu_A = \mu_B$

4. $\sigma_{\bar{X}_1 - \bar{X}} = 6.78$

5. $\sigma_{\bar{X}_1 - \bar{X}_2} = 1.98$

6. $H_0: \mu_1 = \mu_2$; $H_1: \mu_1 \neq \mu_2$

7. *Step 3.* z-statistic.

   Step 4. $\alpha = 0.05$, two-tailed test.

   Step 5. The sampling distribution is the standard normal curve.

   Step 6. Critical region: $z \geq 1.96$ or $z \leq -1.96$

8. $z = 0.69$. Since z is not in the critical region, we fail to reject H_0.

9. $t = \dfrac{\bar{X}_1 - \bar{X}_2 - 12}{s_{\bar{X}_1 - \bar{X}_2}}$

10. $t = \dfrac{\bar{X}_2 - \bar{X}_1 - 7}{s_{\bar{X}_1 - \bar{X}_2}}$

11. $t = \dfrac{\bar{X}_A - \bar{X}_B}{s_{\bar{X}_1 - \bar{X}_2}}$

12. $s_{\bar{X}_1 - \bar{X}_2} = 2.15$

13. $s_{\bar{X}_1 - \bar{X}_2} = 1.35$

14. $s_{\bar{X}_1 - \bar{X}_2} = 0.99$

15. $s_{\bar{X}_1 - \bar{X}_2} = 2.06$

16. $H_0: \mu_1 = \mu_2$; $H_1: \mu_1 \neq \mu_2$

17. *Step 3.* The appropriate test statistic is the Student t-ratio.

    Step 4. $\alpha = 0.01$.

    Step 5. The Student t-distribution with 22 df.

    Step 6. The critical region consists of all values of $t \geq 2.819$ and $t \leq -2.819$.

18. *Step 7.* $t = 0.97$; since obtained t is not in the critical region, we fail to reject H_0.

20. *Step 1.* $H_0: \mu_1 \leq \mu_2$

    Step 2. $H_1: \mu_1 > \mu_2$

    Step 6. Critical region: $t \geq 1.717$. Fail to reject H_0.

19. $t = 0.615$, df $= 22$. Fail to reject H_0.

21. $F = 2.15$

22. $F = 3.00$

23. $F = 3.67$

24. $F = 4.80$

25. Since the critical value of F is equal to or greater than 3.41, we fail to reject H_0. The hypothesis of homogeneity of variances is tenable.

26. Since the critical value of F is equal to or greater than 3.06, obtained F is in the critical region. Reject hypothesis of homogeneity of variance.

27. Since the critical value of F is equal to or greater than 3.12, obtained F is in the critical region. Null hypothesis of homogeneity of variance is not tenable.

28. Since the critical value of F is equal to or greater than 5.05, obtained F is not in the critical region. Fail to reject H_0.

29. 19

30. 12

31. $F = 1.75$, df $= 10/8$. Since the critical value of F is equal to or greater than 4.30, obtained F is not in the critical region. We fail to reject H_0.

32. $\omega^2 = 0.00$

33. $\omega^2 = 0.00$

34. $\omega^2 = 0.29$

35. $\omega^2 = 0.11$

CHAPTER 4 TEST

1. d   2. a   3. b   4. a

5. c   6. d   7. c   8. d

9. d   10. c

11. $\sigma_{X_1 - X_2} = \sqrt{\dfrac{108}{30} + \dfrac{110}{33}} = \sqrt{63.00} = 7.94$

$z = \dfrac{3.06}{7.94} = 0.39$

The critical region for rejecting the nondirectional null hypothesis at $\alpha = 0.01$ is a z of ± 2.58. Since obtained $z = 0.39$ is less than this value, we fail to reject H_0.

13. $s_1^2 = \dfrac{244.89}{n-1} = 30.61$

$s_2^2 = \dfrac{119.88}{n-1} = 17.12$

$F = \dfrac{30.61}{17.12},$ df 9/8

     $= 1.79$

The critical value for rejecting H_0 is 4.36. Since obtained F is less than the critical value, we fail to reject H_0. We may treat the variance as homogeneous.

14. a) $\omega^2 = \dfrac{6.92 - 1}{6.92 + 9 - 1} = 0.40$

b) $\omega^2 = \dfrac{6.92 - 1}{6.92 + 18 - 1} = 0.25$

c) $\omega^2 = \dfrac{6.92 - 1}{6.92 + 27 - 1} = 0.18$

12. $\sum X_1^2 = 2037, \sum X_1 = 127$

$\bar{X}_1 = 14.11, n_1 = 9$

$\sum x_1^2 = 2037 - \dfrac{(127)^2}{9} = 244.89$

$\sum X_2^2 = 445, \sum X_2 = 51$

$\bar{X}_2 = 6.38, n_2 = 8$

$\sum x_2^2 = 445 - \dfrac{(51)^2}{8} = 119.88$

$t = \dfrac{14.11 - 6.38}{\sqrt{\dfrac{244.89 + 119.88}{15}\left(\dfrac{1}{9} + \dfrac{1}{8}\right)}} = \dfrac{7.73}{\sqrt{\dfrac{28.29}{15}}}$

$= \dfrac{7.73}{1.37} = 5.64$

The critical value of t at 15 df for rejecting directional hypothesis at $\alpha = 0.01$ is 2.602. Since obtained t exceeds this value, we reject H_0 and assert H_1.

CHAPTER 5 EXERCISES

1. $0.151; 0.234$

2. $0.590; 0.793$

3. $0.192; 0.110$

4. $0.366; 0.060$

5. 0.24

6. 0.18

7. 0.13

8. 0.13

9. $z = -0.35$; since the two-tailed probability of $z = -0.35$ is 0.73 and this is greater than 0.05, we cannot reject H_0.

10. $z = 0.83$; since the two-tailed probability of $z = 0.83$ is 0.41 and this is greater than 0.05, we cannot reject H_0.

11. $z = 0.63$; since the two-tailed probability of $z = 0.63$ is 0.53 and this is greater than 0.05, we cannot reject H_0.

12. $z = 3.28$; since the two-tailed probability of $z = 3.28$ is approximately 0.001 and this is less than 0.05, we reject H_0 and assert that the correlation coefficients are significantly different.

13. Same as 9

14. Same as 10

15. Same as 11

16. Same as 12

CHAPTER 5 TEST

1. c

2. b

3. d

4. a

5. a

6. d

7. c

8. a

9. $s_{D_z} = \sqrt{\dfrac{1}{157} + \dfrac{1}{120}} = 0.12$

$z = \dfrac{0.096 - 0.400}{0.12} = 4.97$

The critical value of z at $\alpha = 0.05$, two-tailed test, is ± 1.96. Since obtained z is more extreme than 1.96, we reject H_0 and assert H_1.

10. $s_{D_z} = \sqrt{\dfrac{1}{51} + \dfrac{1}{60}} = 0.19$

$z = \dfrac{0.277 - (-0.181)}{0.19} = 2.41$

The critical value of z at $\alpha = 0.01$, one-tailed test, is 2.33. Since obtained z is more extreme than 2.33, we reject H_0 and assert H_1.

11. $s_{D_z} = \dfrac{1}{413} + \dfrac{1}{402} = 0.07$

$z = \dfrac{0.245 - 0.070}{0.07} = 2.50$

The critical value of z at $\alpha = 0.01$, two-tailed test, is 2.58. Since obtained z is less extreme than the critical value, we fail to reject H_0.

CHAPTER 6 EXERCISES

1. Before-after design

2. Matched group design

3. $t = \dfrac{1.93}{2.01} = 0.96$

4. $t = \dfrac{2.06}{2.04} = 1.01$

5. $t = -\dfrac{5.73}{3.91} = -1.47$

6. $t = \dfrac{1.89}{0.58} = 3.26$

7.

X_1	X_2	D	D^2
15	18	-3	9
12	14	-2	4
11	13	-2	4
9	9	0	0
7	6	1	1
		$\sum D = 6$	$\sum D^2 = 18$

$\sum d^2 = \sum D^2 - \dfrac{(\sum D)^2}{n} = 18 - 7.2 = 10.8$

$s_{\bar{D}} = \sqrt{\dfrac{\sum d^2}{n(n-1)}} = \sqrt{\dfrac{10.8}{20}} = 0.73$

8.

X_1	X_2	D	D^2
25	22	3	9
21	19	2	4
15	18	-3	9
14	11	3	9
12	9	3	9
		$\sum D = 8$	$\sum D^2 = 40$

$\sum d^2 = 40 - \dfrac{(8)^2}{5} = 27.2$

$s_{\bar{D}} = \sqrt{\dfrac{27.2}{20}} = 1.17$

9. $\bar{D} = 1.20$

10. $\bar{D} = 1.60$

11.

Item	Grocery	Supermarket	D	D^2
1	0.98	0.86	0.12	0.0144
2	0.23	0.23	0.00	0.0000
3	0.18	0.18	0.00	0.0000
4	0.42	0.39	0.03	0.0009
5	0.57	0.63	-0.05	0.0025
6	0.48	0.49	-0.01	0.0001
7	0.88	0.79	0.09	0.0081
8	1.33	1.29	0.04	0.0016
9	1.82	1.73	0.08	0.0064
10	1.11	0.99	0.12	0.0144
			$\sum D = 0.37$	$\sum D^2 = 0.0484$

$\sum d^2 = 0.0408 - \dfrac{(0.37)^2}{10} = 0.0347$

$s_{\bar{D}} = \sqrt{\dfrac{0.0347}{90}} = 0.0196$

$\bar{D} = \dfrac{0.37}{10} = 0.037$

$t = \dfrac{0.037}{0.0196} = 1.89, \quad df = 9$

Since obtained t does not exceed the critical value of 2.228, we cannot reject H_0.

12. $A = \dfrac{120}{(5)^2} = 4.80$ 13. $A = \dfrac{120}{(40)^2} = 0.075$ 14. $A = \dfrac{120}{(40)^2} = 0.1875$

15. The critical value of A at $\alpha = 0.01$, one-tailed test, with df $= 40$ is 0.158. Since obtained A exceeds this value, we cannot reject H_0.

16. The critical value of A at $\alpha = 0.01$, two-tailed test, with df $= 30$ is 0.160. Since the obtained A is less than this value, it is in the critical region for rejecting H_0. We therefore reject H_0 and assert H_1.

17. The critical value of A at $\alpha = 0.05$, two-tailed test, with df $= 17$ is 0.268. Since obtained A is less than this value, we reject H_0 and assert H_1.

18.

Experimental	Control	D	D^2
25	21	4	16
23	20	3	9
21	24	-3	9
19	15	4	16
17	20	-3	9
17	15	2	4
16	14	2	4
14	10	4	16
14	12	2	4
12	13	-1	1
11	6	5	25
10	9	1	1
		$\sum D = 20$	$\sum D^2 = 114$

1. H_0: There is no effect of the drug on the task of motor coordination; that is, $\mu_D = 0$.

2. H_1: There is an effect of the drug on the motor coordination task; that is, $\mu_D \neq 0$.

3. *Statistical test*: Since we are employing a matched group design, the Sandler A-statistic for correlated samples is appropriate.

4. *Significance level*: $\alpha = 0.05$.

5. *Sampling distribution*: The Sandler A-statistic with df $= 11$.

6. *Critical region*: $A_{0.05} \leq 0.273$

$$A = \frac{114}{(20)^2} = 0.285$$

Since obtained A exceeds the critical value, we cannot reject H_0.

CHAPTER 6 TEST

1. a   2. c   3. d   4. c

5. b   6. d   7. d   8. b

9. a   10. a

11.

Experimental condition	Control condition	D	D²
106	110	−4	16
103	107	−4	16
100	105	−5	25
99	94	5	25
96	99	−3	9
94	101	−7	49
91	93	−2	4
90	94	−4	16
87	90	−3	9
84	82	2	4
82	87	−5	25
78	84	−6	36
		$\sum D = -36$	$\sum D^2 = 234$

$$\sum d^2 = 234 - \frac{(36)^2}{12} = 126$$

$$s_{\bar{D}} = \sqrt{\frac{126}{132}} = 0.98$$

$$\bar{D} = -3.00$$

$$t = -\frac{3.00}{0.98} = -3.06, \qquad df = 11$$

The critical value of t for a nondirectional hypothesis at $\alpha = 0.05$ is -1.796. Since obtained t is more extreme than this value, it is in the critical region. We reject H_0 and assert H_1.

$$A = \frac{234}{(-36)^2} = \frac{234}{1296}$$

$$= 0.1806, \qquad df = 11$$

The critical value of A for a nondirectional hypothesis at $\alpha = 0.05$ is 0.368. Since obtained A is less than this value, we reject H_0 and assert H_1.

12. $t = \sqrt{\dfrac{11}{2.1672 - 1}} = 3.07$

The slight disparity of 0.01 represents rounding error.

CHAPTER 7 EXERCISES

1. On the basis of the data, we would reject H_0. However, since H_0 is true, we have made a type I or type α error.

2. On the basis of the data, we would fail to reject H_0. Since H_0 is true, we have made a correct decision.

3. We fail to reject H_0. Since H_0 is false, we have failed to reject a false H_0. Therefore, we have made a type II or type β error.

4. We reject H_0. Since H_0 is false, we have made a correct decision.

5. We reject H_0. Since H_0 is false, we have made a correct decision.

6. The probability of making a type I error is equal to α. In the present case, $\alpha = 0.05$.

7. a) *Null hypothesis*: $\mu_1 \leq \mu_2$.

   b) *Alternative hypothesis*: $\mu_1 > \mu_2$.

   c) *Statistical test*: Since σ_1 and σ_2 are known, the appropriate test statistic is

   $$z = \frac{\bar{X}_1 - \bar{X}_2}{\sigma_{\bar{X}_1 - \bar{X}_2}}.$$

   d) *Significance level*: $\alpha = 0.01$, one-tailed test.

   e) *Sampling distribution*: the normal curve.

   f) *Critical region*: $z \geq 2.33$; the critical value of the difference in sample means is $\bar{X}_1 - \bar{X}_2 = 2.33\sigma_{\bar{X}_1 - \bar{X}_2}$.

8. $\sigma_{\bar{X}_1 - \bar{X}_2} = \sqrt{\dfrac{15}{20} + \dfrac{15}{20}} = 1.22$

9. Critical value: $\bar{X}_1 - \bar{X}_2 = (2.33)(1.22) = 2.84$.

10. The z corresponding to the critical value in the true sampling distribution under H_1 is

$$z = \frac{2.84 - 5.00}{1.22} = -\frac{2.16}{1.22} = -1.77.$$

11. The area beyond $z = -1.77$ is 0.0384. Thus the probability of a type II error is 0.0384.

12. Power $= 1 - 0.0384 = 0.9616$

13. $\sigma_{\bar{X}_1 - \bar{X}_2} = \sqrt{\dfrac{20}{10} + \dfrac{20}{10}} = 2.00$

Critical value: $\bar{X}_1 - \bar{X}_2 = \pm 1.96(2.00) = \pm 3.92$

$$z = \frac{-3.92 - 4.00}{2.00} = \frac{-7.92}{2.00} = -3.96$$

The associated probability is so small that we shall ignore this possibility.

$$z = \frac{3.92 - 4.00}{2.00} = -0.04 \qquad \text{Power} = 0.5160$$

14. $\sigma_{\bar{X}_1 - \bar{X}_2} = \sqrt{\dfrac{20}{40} + \dfrac{20}{40}} = 1.00$

Critical value: $\bar{X}_1 - \bar{X}_2 = \pm 1.96$

$$z = \frac{-1.96 - 4.0}{1.00} = -5.96$$

The associated probability is so small that we shall ignore this possibility.

$$z = \frac{1.96 - 4.00}{1.00} = -2.04$$

Power $= 1 - 0.0207 = 0.9793$

16. As you increase sample size, you increase the power of the test. However, the increase is not linear. There is a declining *rate of increase* of power with increased N.

17. Power $= 0.7257$ when $\alpha = 0.01$.

15.

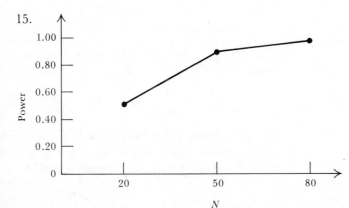

18. $\sigma_{\bar{X}_1 - \bar{X}_2} = 1.26$

    Critical value: $\bar{X}_1 - \bar{X}_2 = (\pm 2.58)(1.26) = \pm 3.25$

    $z = \dfrac{-3.25 - 4.00}{1.26} = \dfrac{-7.25}{1.26} = -5.75$

    The associated probability is so small that we shall ignore this possibility.

    $z = \dfrac{3.25 - 4.00}{1.26} = \dfrac{-.75}{1.26} = -0.60$

    Power $= 1 - 0.2257 = 0.7743$

19. c) $r = 0.81$

CHAPTER 7 TEST

1. d	2. b	3. a	4. c
5. a	6. c	7. d	8. d
9. b	10. b		

11. a) $\sigma_{\bar{X}_1 - \bar{X}_2} = \sqrt{\dfrac{9}{40} + \dfrac{9}{40}} = 0.67$

    Critical value $\bar{X}_1 - \bar{X}_2 = \pm 1.96(0.67) = \pm 1.31$

    $z = \dfrac{-1.31 - 3}{0.67} = \dfrac{-4.31}{0.67} = -6.43$

    The probability is so small that we shall ignore the possibility.

    $z = \dfrac{1.31 - 3}{0.67} = \dfrac{-1.69}{0.67} = -2.52$

    The probability of obtaining in the true distribution a difference in sample means such that $z \le -2.52$ is 0.0059. Thus the power of the test is 0.9941.

    b) $\sigma_{\bar{X}_1 - \bar{X}_2} = \sqrt{\dfrac{9}{20} + \dfrac{9}{20}} = 0.95$

    Critical value $\bar{X}_1 - \bar{X}_2 = \pm 2.58(0.95) = \pm 2.45$

    $z = \dfrac{-2.45 - 3}{0.95} = \dfrac{-5.45}{0.95} = -5.74$

    The probability is so small that we shall ignore the possibility.

    $z = \dfrac{2.45 - 3}{0.95} = \dfrac{-0.55}{0.95} = -0.58$

    The probability of obtaining in the true distribution a difference in sample means such that $z \le -0.58$ is 0.2810. The power of the test is therefore 0.7190.

    c) $\sigma_{\bar{X}_1 - \bar{X}_2} = 0.95$

    Critical value $\bar{X}_1 - \bar{X}_2 = \pm 1.96(0.95) = \pm 1.86$

    $z = \dfrac{-1.86 - 3}{0.95} = \dfrac{-4.86}{0.95} = -5.12$

    The probability is so small that we shall ignore the possibility.

    $z = \dfrac{1.86 - 3}{0.95} = \dfrac{-1.14}{0.95} = -1.20$

    The probability of obtaining in the true distribution a difference in sample means such that $z \le -1.20$ is 0.1151. Therefore, the power of the test is 0.8849.

    d) $\sigma_{\bar{X}_1 - \bar{X}_2} = 0.67$

    Critical value $\bar{X}_1 - \bar{X}_2 = \pm 2.58(0.67) = \pm 1.73$

    $z = \dfrac{-1.73 - 3}{0.67} = \dfrac{-4.73}{0.67} = -7.06$

    The probability is so small that we shall ignore the possibility.

    $z = \dfrac{1.73 - 3}{0.67} = \dfrac{-1.27}{0.67} = -1.90$

    The probability of obtaining in the true distribution a difference in sample means such that $z \le -1.90$ is 0.0287. Thus the power of the test is 0.9713.

CHAPTER 8 EXERCISES

1. Quantitative

2. Quantitative

3. Quantitative

4. Quantitative

5. Quantitative

6. H_0: All eight treatment groups were drawn from a common population of means.
   H_1: All eight treatment groups were not drawn from a common population of means.

7. H_0: The four treatment groups were selected from the same population of means.
   H_1: The four treatment groups were not selected from the same population of means.

8. Small

9. Large

10. Large

11. 2.75

12. 1.00

13. 1.88

14. 1.00

15. 5.00

16. $\sum x_T^2 = 202.5$

17. $\sum x_W^2 = 180.0$

18. $\sum x_B^2 = 22.5$

19. $202.5 = 180.0 + 22.5 = 202.5$

20. a) $df_T = N - 1 = 9$
    b) $df_W = N - k = 8$
    c) $df_B = k - 1 = 1$

21. a) $s_W^2 = \dfrac{180.0}{8} = 22.5$
    b) $s_B^2 = \dfrac{22.5}{1} = 22.5$

22.

Source of variation	Sum of squares	Degrees of freedom	Variance estimate	F
Between groups	8.00	1	8.00	1.20
Within groups (error)	40.00	6	6.67	
Total	48	7		

F required for significance at $\alpha = 0.05$ is 5.99. Since obtained F is less than the critical value, we fail to reject H_0.

23.

Source of variation	Sum of squares	Degrees of freedom	Variance estimate	F
Between groups	611	4	152.75	7.91
Within groups (error)	483	25	19.32	
Total	1094	29		

F required for significance at $\alpha = 0.01$ is 4.18. Since obtained F exceeds the critical value, it is in the critical region. Reject H_0 and assert alternative hypothesis.

CHAPTER 8 TEST

1. a   2. b   3. b   4. a

5. c   6. d   7. c   8. a

9. c   10. b   11. d   12. a

13. c   14. a   15. b

16. *Step 1.* $\sum X_T = 37 + 46 + 49 + 50 = 182$ *Step 2.* $\dfrac{(\sum X_T)^2}{N} = \dfrac{(182)^2}{32} = 1035.12$

Step 3. $\sum X_T^2 = 233 + 302 + 363 + 360 = 1258$ *Step 4.* $\sum X_T^2 = 1258 - 1035.12 = 222.88$

Step 5. $df_T = 31$ *Step 6.* See summary table at Step 13.

Step 7. $\sum x_B^2 = \dfrac{(37)^2 + (46)^2 + (49)^2 + (50)^2}{8}$ *Step 8.* $df_B = 3$

$\qquad\qquad - 1035.12$ *Step 9.* See summary table at Step 13.

$\qquad\quad = 1048.25 - 1035.12 = 13.13$

Step 10. $\sum x_W^2 = \left[233 - \dfrac{(37)^2}{8}\right] + \left[302 - \dfrac{(46)^2}{8}\right]$ *Step 11.* $df_W = 28$

$\qquad\qquad + \left[363 - \dfrac{(49)^2}{8}\right] + \left[360 - \dfrac{(50)^2}{8}\right]$ *Step 12.* See summary table at Step 13.

$\qquad\quad = 61.88 + 37.50 + 62.88 + 47.50$

$\qquad\quad = 209.75$

Step 13.

Source of variation	Sum of squares	Degrees of freedom	Variance estimate	*F*
Between groups	13.13	3	4.38	0.58
Within groups	209.75	28	7.49	
Total	222.88	31		

Step 14. $\sum x_T^2 = \sum x_B^2 + \sum x_W^2$ *Step 15.* $s_B^2 = \dfrac{13.13}{3} = 4.38$

$\qquad\quad = 13.13 + 209.75$

$\qquad\quad = 222.88$ *Step 16.* $s_W^2 = \dfrac{209.75}{28} = 7.49$

Step 17. $F = \dfrac{4.38}{7.49} = 0.58,\qquad df\ 3/28$ *Step 18.* The critical value of F at 3 and 28 degrees of freedom, $\alpha = 0.01$, is 4.57. Since obtained F falls short of this value, we fail to reject H_0.

CHAPTER 9 EXERCISES

1. Unplanned comparison   2. Planned comparisons   3. 3

4. 10   5. 15   6. 21   7. 1

8.

	$\bar{X}_1 = 20.00$	$\bar{X}_2 = 15.83$	$\bar{X}_3 = 12.33$	$\bar{X}_4 = 9.50$	$\bar{X}_5 = 7.33$
$\bar{X}_1 = 20.00$	—	4.17	7.67	10.50	12.67
$\bar{X}_2 = 15.83$	—	—	3.50	6.33	8.50
$\bar{X}_3 = 12.33$	—	—	—	2.83	5.00
$\bar{X}_4 = 9.50$	—	—	—	—	2.17
$\bar{X}_5 = 7.33$	—	—	—	—	—

$$HSD = 4.16^* \sqrt{\frac{19.32}{6}} = (4.16)(1.79) = 7.34$$

The differences between $\bar{X}_1$, and $\bar{X}_3$, $\bar{X}_4$, and $\bar{X}_5$ are statistically significant. The difference between $\bar{X}_2$ and $\bar{X}_5$ is statistically significant.

CHAPTER 9 TEST

1. d   2. a   3. b   4. c

5. a

6. $\bar{X}_1 = 14.14$, $\bar{X}_2 = 12.99$, $\bar{X}_3 = 15.78$

Step 1.

Step 2. $q_\alpha = 3.53$

	$\bar{X}_1 = 14.14$	$\bar{X}_2 = 12.99$	$\bar{X}_3 = 15.78$
$\bar{X}_1 = 14.14$	—	−1.15	1.64
$\bar{X}_2 = 12.99$	—	—	2.79
$\bar{X}_3 = 15.78$	—	—	—

Step 3. HSD equals $3.53 \dfrac{1.90}{9} = (3.53)(0.46) = 1.62$

Step 4. The following differences are statistically significant at $\alpha = 0.05$: $\bar{X}_1$ versus $\bar{X}_3$, $\bar{X}_2$ versus $\bar{X}_3$.

---

* Obtained by linear interpolation: $4.10 + \frac{6}{7}(4.17 - 4.10) = 4.16$.

CHAPTER 10 EXERCISES

1. The following subjects should be in each block:

   Block 1   A   B   L
   Block 2   C   D   K
   Block 3   E   F   H
   Block 4   G   I   J

2. The following subjects should be in each block:

   Block 1   A   C   I   K
   Block 2   B   D   E   H
   Block 3   F   G   J   L

3. $\sum X_T = 295$

4. $(\sum X_T)^2 = 87{,}025$

5. $\sum X_T^2 = 4757$

6. $\sum x_T^2 = 4757 - \dfrac{87{,}025}{20} = 4757 - 4351.25 = 405.75$

7. $df_T = N - 1 = 19$

8. See Exercise 18.

9. $\sum x_B^2 = \dfrac{(60)^2 + (69)^2 + (77)^2 + (89)^2}{5} - 4351.25$

10. $df_B = k - 1 = 3$

$$= \dfrac{22{,}211}{5} - 4351.25 = 4442.20 - 4351.25 = 90.95$$

11. See Exercise 18.

12. $\sum x_{bl}^2 = \dfrac{(72)^2 + (76)^2 + (60)^2 + (52)^2 + (35)^2}{4} - 4351.25$

$$= \dfrac{18{,}489}{4} - 4351.25 = 4622.25 - 4351.25 = 271.00$$

13. $df_{bl} = bl - 1 = 4$

14. See Exercise 18.

15. $\sum x_{bl \times B}^2 = \sum x_T^2 - (\sum x_B^2 + \sum x_{bl}^2)$
$$= 405.75 - (90.95 + 271.00) = 43.8$$

16. $df_{bl \times B} = (bl - 1)(B - 1)$
$$= (4)(3) = 12$$

17. See Exercise 18.

18.

Source of variation	Sum of squares	Degrees of freedom	Variance estimate	F
Between groups	90.95	3	30.32	8.31
Between blocks	271.00	4		
Residual (error)	43.8	12	3.65	
Total	405.75	20		

19. $90.95 + 271.00 + 43.8 = 405.75$

    $3 + 4 + 12 = 20$

20. $s_B^2 = 30.32$

21. $s_{bl \times B}^2 = 3.65$

22. $F = 8.31$, df $= 3/12$

23. The critical value of F at $\alpha = 0.05$ and $3/12$ degrees of freedom is 3.49. Since obtained F exceeds this value, it is in the critical region. We must reject H_0 and assert H_1.

CHAPTER 10 TEST

1. b	2. b	3. a	4. a
5. d	6. c	7. d	8. c
9. a	10. b	11. d	12. a

13. The blocks would contain the following subjects, not necessarily in the order shown:

Block 1: D, L, F, J Block 2: O, B, N, K
Block 3: F, H, P, C Block 4: G, E, A, M

14. *Step 1.* $\sum X_T = 98 + 76 + 100 = 274$ *Step 2.* $\dfrac{(\sum X_T)^2}{N} = \dfrac{(274)^2}{18} = 4170.89$

Step 3. $\sum X_T^2 = 1844 + 1110 + 1828 = 4782$ *Step 4.* $\sum x_T^2 = 4782 - 4170.89 = 611.11$

Step 5. $df_T = 18 - 1 = 17$ *Step 6.* See summary table at Step 16.

Step 7. $\sum x_B^2 = \dfrac{(98)^2 + (76)^2 + (100)^2}{6} - 4170.89 = 59.11$

Step 8. $df_B = 3 - 1 = 2$ *Step 9.* See summary table at Step 16.

Step 10. $\sum x_{bl}^2 = \dfrac{(67)^2 + (571)^2 + (571)^2 + (421)^2 + (271)^2 + (24)^2}{3} - 4170.89 = 514.44$

Step 11. $df_{bl} = 6 - 1 = 5$ *Step 12.* See summary table at Step 16.

Step 13. $\sum x_{bl \times B}^2 = 611.11 - (59.11 + 514.44) = 37.56$

Step 14. $df_{bl \times B} = 2 \times 5 = 10$ *Step 15.* See summary table at Step 16.

Step 16.

Source of variation	Sum of squares	Degrees of freedom	Variance estimate	F
Between groups	59.11	2	29.56	7.86
Between blocks	514.44	5	—	
Residual (error)	37.56	10	3.76	
Total	611.11	17		

Step 17. $\sum x_T^2 = \sum x_B^2 + \sum x_{bl}^2 + \sum x_{bl \times B}^2$

$611.11 = 59.11 + 514.44 + 37.56$

$df_T = df_B + df_{bl} + df_{bl \times B}$

$= 2 + 5 + 10 = 17$

Step 18. $s_B^2 = \dfrac{59.11}{2} = 29.56$

Step 19. $s_{bl \times B}^2 = \dfrac{57.56}{10} = 3.76$

Step 20. $F = \dfrac{29.56}{3.76} = 7.86$

Step 21. The critical value of F at $\alpha = 0.01$ and 2 and 10 degrees of freedom is 7.56. Since obtained F exceeds this value, reject H_0 and assert H_1.

CHAPTER 11 EXERCISES

1. 4×2 2. 3×5 3. 2×2 4. 5×7

5. 5×7 6.

	A_1		A_2		A_3		A_4	
	B_1	B_2	B_1	B_2	B_1	B_2	B_1	B_2
Treatment combinations	A_1B_1	A_1B_2	A_2B_1	A_2B_2	A_3B_1	A_3B_2	A_4B_1	A_4B_2

7.

	A_1			A_2			A_3		
	B_1	B_2	B_3	B_1	B_2	B_3	B_1	B_2	B_3
Treatment combinations	A_1B_1	A_1B_2	A_1B_3	A_2B_1	A_2B_2	A_2B_3	A_3B_1	A_3B_2	A_3B_3

8.

	A_1			A_2			A_3			A_4			A_5		
	B_1	B_2	B_3	B_1	B_2	B_3	B_1	B_2	B_3	B_1	B_2	B_3	B_1	B_2	B_3
Treatment combina- tions	A_1B_1	A_1B_2	A_1B_3	A_2B_1	A_2B_2	A_2B_3	A_3B_1	A_3B_2	A_3B_3	A_4B_1	A_4B_2	A_4B_3	A_5B_1	A_5B_2	A_5B_3

9. 12   10. 9   11. 12   12. 20

13. 60   14. $\sum x_T^2 = \sum x_W^2 + \sum x_{TC}^2$ 15. $\sum x_{TC}^2 = \sum x_C^2 + \sum x_D^2 + \sum x_{C \times D}^2$

16. $s_C^2 = \dfrac{\sum x_C^2}{df_C}$ 17. $s_W^2 = \dfrac{\sum x_W^2}{df_W}$

$s_D^2 = \dfrac{\sum x_D^2}{df_D}$ 18. $\sum x_T^2 = 1470 - \dfrac{(176)^2}{24} = 1470 - 1290.67 = 179.33$

$s_{C \times D}^2 = \dfrac{\sum x_{C \times D}}{df_{C \times D}}$ $df_T = N - 1 = 23$

19. $\sum x_{TC}^2 = \dfrac{(26)^2 + (23)^2 + (17)^2 + (35)^2 + (35)^2 + (40)^2}{4} - 1290.67$

$= \dfrac{5544}{4} - 1290.67 = 95.33$

$df_{TC} = TC - 1 = 5$

20. $\sum x_A^2 = \dfrac{(66)^2 + (110)^2}{12} - 1290.67$ 21. $\sum x_B^2 = \dfrac{(61)^2 + (58)^2 + (57)^2}{8} - 1290.67$

$= \dfrac{16,456}{12} - 1290.67 = 80.66$ $= \dfrac{10,334}{8} - 1290.67 = 1.13$

$df_A = A - 1 = 1$ $df_B = B - 1 = 2$

22. $\sum x_{A \times B}^2 = 95.33 - (80.66 + 1.13) = 13.54$

    $\mathrm{df}_{A \times B} = (A - 1)(B - 1) = 2$

23. $\sum x_W^2 = 84.00$

    $\mathrm{df}_W = \mathcal{N} - TC = 18$

24.

Source of variation	Sum of squares	Degrees of freedom	Variance estimate	F
Treatment combinations	95.33	5		
$\quad$ A variable	80.66	1	80.66	17.31
$\quad$ B variable	1.13	2	0.56	0.12
$\quad$ A × B	13.54	2	6.77	1.45
Within (error)	84.00	18	4.66	
$\qquad$ Total	179.33	23		

25. $\sum x_T^2 = \sum z_{TC}^2 + \sum x_W^2$

    $179.33 = 95.33 + 84.00$

    $179.33 = 179.33$

    $\sum x_{TC}^2 = \sum x_A^2 + \sum x_B^2 + \sum x_{A \times B}^2$

    $95.33 = 80.66 + 1.13 + 13.54$

    $95.33 = 95.33$

26. $s_{A \times B}^2 = 1.45$

28. $s_A^2 = 17.31$

27. $s_B^2 = 0.12$

29. $s_W^2 = 4.66$

30. $F_{A \times B} = 1.45$, df 2/18. Critical value: $F \geq 3.55$. Fail to reject H_0.

31. $F_B = 0.12$, df 2/18. Critical value: $F \geq 3.55$. Fail to reject H_0.

32. $F_A = 17.31$, df 1/18. Critical value: $F \geq 4.41$. Since $F_A \geq 4.41$, reject H_0 and assert H_1.

33. Tukey HSD test with A variable is not necessary since there are only two conditions compared. The fact that F is significant for the A condition means that the two treatment levels are significantly different. No follow-up analysis on the B variable and the $A \times B$ combination is warranted since these conditions did not achieve statistical significance.

CHAPTER 11 TEST

1. c

2. b

3. a

4. d

5. c

6. b

7. c

8. a

9. d

10. a

11. c

12. b

13. a

14. c

15. c

16. *Step 1.* $\sum x_T^2 = 5274 - 4629.63 = 644.37$

    Step 3. See summary table at Step 19.

    Step 2. $\mathrm{df}_T = 54 - 1 = 53$

    Step 4. $\sum x_{TC}^2 = \dfrac{(41)^2 + (55)^2 + \cdots + (70)^2}{6} - 4629.63$

    $= 4798.67 - 4629.63 = 169.04$

    Step 5. $\mathrm{df}_{TC} = 9 - 1 = 8$

    Step 6. See summary table at Step 19.

Step 7. $\quad \sum x_A^2 = \dfrac{(163)^2 + (161)^2 + (176)^2}{18} - 4629.63$

$\quad\quad\quad\quad = 4637 - 4629.63 = 7.37$

Step 8. $\quad \mathrm{df}_A = 3 - 1 = 2$

Step 9. See summary table at Step 19.

Step 10. $\quad \sum x_B^2 = \dfrac{(127)^2 + (170)^2 + (203)^2}{18} - 4629.63$

$\quad\quad\quad\quad = 4791 - 4629.63 = 161.37$

Step 11. $\quad \mathrm{df}_B = 3 - 1 = 2$

Step 12. See summary table at Step 19.

Step 13. $\quad \sum x_{A \times B}^2 = 169.04 - (7.37 + 161.37)$

$\quad\quad\quad\quad = 0.30$

Step 14. $\quad \mathrm{df}_{A \times B} = 2 \times 2 = 4$

Step 15. See summary table at Step 19.

Step 16. $\quad \sum x_W^2 = 644.37 - 169.04 = 475.33$

Step 17. $\quad \mathrm{df}_W = 54 - 9 = 45$

Step 18. See summary table at Step 19.

Step 19.

Source of variation	Sum of squares	Degrees of freedom	Variance estimate	F
Treatment combinations	169.04	8		
$\quad$ *A*-variable	7.37	2	3.68	0.35
$\quad$ *B*-variable	161.37	2	80.68	7.64
$\quad$ *A* × *B*	0.30	4	0.08	0.01
Within (error)	475.33	45	10.56	
$\quad$ Total	644.37	53		

Step 20. $\quad \sum x_{TC}^2 + \sum x_W^2 = \sum x_T^2$

$\quad\quad\quad 169.04 + 475.33 = 644.37$

$\quad\quad\quad \sum x_A^2 + \sum x_B^2 + \sum x_{A \times B}^2 = \sum x_{TC}^2$

$\quad\quad\quad 7.37 + 161.37 + 0.30 = 169.04$

Step 21. $\quad s_{A \times B}^2 = \dfrac{0.30}{4} = 0.08$

Step 22. $\quad s_B^2 = \dfrac{161.37}{2} = 80.68$

Step 23. $\quad s_A^2 = \dfrac{7.37}{2} = 3.68$

Step 24. $\quad s_W^2 = \dfrac{475.33}{45} = 10.56$

Steps 25, 26. $\quad F_{A \times B} = \dfrac{0.08}{10.56}$

$\quad\quad\quad\quad = 0.01$, df 4/45, not significant.

Steps 27, 28. $\quad F_B = \dfrac{80.68}{10.56}$

$\quad\quad\quad\quad = 7.64$, df 2/45, significant.

Steps 29, 30. $\quad F_A = \dfrac{7.08}{10.56}$

$\quad\quad\quad\quad = 0.35$, df 2/45, not significant.

17. Since the *B*-variable yields statistical significance, we apply the Tukey HSD Test:

$$\bar{X}_{B_1} = \frac{127}{18} = 7.06, \quad \bar{X}_{B_2} = \frac{170}{18} = 9.44, \quad \bar{X}_{B_3} = \frac{203}{18} = 11.28.$$

Step 1.

	$\bar{X}_{B_1} = 7.06$	$\bar{X}_{B_2} = 9.44$	$\bar{X}_{B_3} = 11.28$
$\bar{X}_{B_1} = 7.06$	—	2.38	4.22
$\bar{X}_{B_2} = 9.44$	—	—	1.84
$\bar{X}_{B_3} = 11.28$	—	—	—

Step 2. $q_\alpha = 0.05 = 3.43$, by linear interpolation.   *Step 3.*   HSD $= 3.43 \sqrt{\dfrac{10.56}{18}} = 2.64$

Step 4.   Only the difference between conditions B_1 and B_3 is significant at the 0.05 level.

Index

A posteriori tests, 90
A priori tests, 90
Alpha level, 14, 16
  and power, 73
Alternative hypothesis, 14–15
Analysis of variance
  one-way, independent samples, 78–86
  one-way, correlated samples, 96–102
  two-way, 106–116
Association, test of, *See* Omega squared
A-statistic, 28–29, 63–65
  table interpretation, 63–64

Between-group sum of squares, 81–82
Between-group variance estimate, 78
Blocks sum of squares, 97

Correlated samples design
  before-after, 58, 96
  matched group, 58, 96
Correlation, *see* Pearson *r*
Critical value, 16–17
  and power, 70–71

Design
  factorial, 107–108
  randomized block, 96–102
Directional hypothesis, 15–16, 54
  and one-tailed tests of significance, 16,
    54
  and power, 74
Distribution
  of mean, 6
  normal, 3–4, 16
  sampling, 5–6
  t, 26–28

Error, precision in estimating related to
    power, 74
Errors, statistical, 68–69
  random error in analysis of variance,
    79
  type α, 68
  type β, 68

Factorial design, 106–116
F-ratio
  and analysis of variance, 79
  and homogeneity of variance, 45
  when H_0 is true, 79
  when H_0 is false, 79

Index

Homogeneity of variance, 45–46
HSD test, 90–92, 102
    calculation of, 91–92
    in factorial design, 116
Hypothesis testing
    alpha level, 16
    alternative hypothesis, 14–15
    correlated samples, 60–61
    critical value, 16
    directional hypothesis, 15–16
    level of significance, 13–14
    nondirectional hypothesis, 15–16
    null hypothesis, 14–15, 60–61
    region of rejection, 16
    significance level, 13–14
    statistical, 12–17
    when parameters are known, 22–24
    when parameters are unknown,
        24–28, 39–46

Interaction sum of squares, 100–101

Multicomparison tests, 89–93
    number of pairwise comparisons,
        90–97
    planned (*a priori*), 90
    unplanned (*a posteriori*), 90

Nondirectional hypothesis, 15–16, 54
    and power, 74
    and two-tailed tests of significance, 16,
        54
Null hypothesis, 14–15
    critical value, 16
    rejection of, 16

Omega squared (ω^2), 46–47
One-sample tests of significance, 22–31
One-tailed tests of significance, 16, 17

Pearson *r*
    independent samples, 52–54
    one-sample case, 30–31
Power of a statistical test, 68, 74
    improving power, 72–74
Probability
    and critical values, 16
    defined, 2
    one-tailed values, 5
    as a proportion, 2

as a proportion of area under the
    normal curve, 3
    and sampling distributions, 5–6
    and standard normal curve, 3–4, 16
    testing significance of Pearson *r*, 31
    and *z*-scores, 3–4
Proportion and probability, 2–3

Random error and *F*-ratio, 79
Randomized block design, 96–102
Region of rejection, 16–17
    and critical value, 16–17
    and standard normal curve, 16–17
Residual sum of squares, 100

Sample size and power, 73
Sampling distributions
    defined, 5–6
    of mean, 7, 12–14
    and power, 69
    and probability, 6–7
    and sample size, 6
    and testing hypothesis, 12–17
Sampling with replacement, 6
Sandler *A*
    and correlated samples, 63–65
    and one-sample case, 28–29
    and relationship to *t*, 29
Standard error of the difference
    between *z*'s, 52
    estimating, 39–41
    n's equal, 40
    n's unequal, 40–41
    when parameters are known, 39
Standard error of the mean
    when parameters known, 22–23
    when parameters unknown, 24–25
Standard error of the mean difference,
    59–60
Standard normal curve and probability, 3
Student *t*-ratio, *see t*-ratio
Sum of squares
    between-group, 81–82
    blocks, 97
    in factorial design, 107–108
    interaction, 100–101
    partitioning, 79–82, 97–102
    raw score formulas, 83–85
    total, 81
    treatment combinations, 107–108
    within-group, 81–82

t-distributions, 26–28
    and normal curve, 26
t-ratio
    correlated samples, 58–63
    one-sample case, 26–28
    and relation to *A*, 29
    two-sample case, 39–46
Testing statistical hypotheses, 12–17
Tests of significance
    one-sample case, 22–31
    one-tailed, 15–16
    Pearson *r*, independent samples,
        52–54
    Pearson *r*, one-sample case, 30–31
    Sandler *A*, 28–29
    t-ratio, 26–28
    two-sample case, correlated samples,
        58–65
    two-sample case, independent
        samples, 36–47
    two-tailed, 15–16
Total sum of squares, 81
Treatment combinations, 106–107
Treatment combinations sum of squares,
    107–108
Treatment effects
    and between-group variance estimate,
        78–79
    and random error in *F*-ratio, 79
Tukey HSD test, 90–92, 102
Two-sample tests of significance, 36–47,
    53
Two-tailed tests of significance, 16, 17
Two-way analysis of variance, 106–116
Type alpha (α) error
    probability of, 68
Type beta (β) error
    calculating probability of, 69–72
Type I error, *see* Type α error
Type II error, *see* Type β error

Variance, analysis of, *see* Analysis of
    variance
Variance estimates
    between-group, 78
    error, 91
    within-group, 78
Variance, homogeneity of, 45–46

Within-group sum of squares, 81–82
Within-group variance estimate, 78

z_r, 30, 52
z-scores and probability, 3–4
z-statistic
    for Pearson r, independent samples,
      53
    for Pearson r, one-sample case, 30–31
    when parameters are known, 22–23,
      26–38